Satellite Remote Sensing for Conservation

Case Studies from Aquatic and Terrestrial Ecosystems

Satellite remote sensing presents an amazing opportunity to inform biodiversity conservation by inexpensively gathering repeated monitoring information for vast areas of the Earth. However, these observations first need processing and interpretation if they are to inform conservation action.

Through a series of case studies, this book presents detailed examples of the application of satellite remote sensing, covering both aquatic and terrestrial ecosystems, to conservation. The authors describe how collaboration between the remote sensing and conservation communities makes satellite data functional for operational conservation, and provide concrete examples of the lessons learned in addition to the scientific details.

The editors, one at the National Aeronautics and Space Administration (NASA) and the other at a conservation non-governmental organisation, have brought together leading researchers in conservation remote sensing to share their experiences from project development through to application, and emphasise the human side of these projects.

ALLISON K. LEIDNER is a conservation biologist and currently works at NASA Headquarters in Washington, DC, for the Earth Science Division. She leads activities within the Biodiversity and Ecological Forecasting Programs, which support the use of remote sensing for basic research and decision-support applications. She also coordinates a variety of climate assessment activities for NASA. Her former research programme focused on the ecology and conservation of rare and endangered species. In 2014, she received a NASA Special Service Award for her efforts related to the development of the Third National Climate Assessment. She is contracted to NASA via ASRC Federal.

GRAEME M. BUCHANAN has a background in field ecology and ornithology, and he has been using remote sensing data for a wide variety of purposes to inform conservation decision-making and prioritisations in the UK and globally for around 15 years. He has served on the British Ornithologist's Union Council, and is an editor for the journals *Bird Conservation International* and *Remote Sensing in Ecology and Conservation*.

Satellite Remote Sensing for Conservation Action

Case Studies from Aquatic and Terrestrial Ecosystems

Edited by

ALLISON K. LEIDNER

*ASRC Federal/National Aeronautics and Space Administration,
Washington DC, USA*

GRAEME M. BUCHANAN

RSPB, Edinburgh, UK

CAMBRIDGE
UNIVERSITY PRESS

CAMBRIDGE
UNIVERSITY PRESS

University Printing House, Cambridge CB2 8BS, United Kingdom

One Liberty Plaza, 20th Floor, New York, NY 10006, USA

477 Williamstown Road, Port Melbourne, VIC 3207, Australia

314–321, 3rd Floor, Plot 3, Splendor Forum, Jasola District Centre, New Delhi – 110025, India

79 Anson Road, #06–04/06, Singapore 079906

Cambridge University Press is part of the University of Cambridge.

It furthers the University's mission by disseminating knowledge in the pursuit of education, learning, and research at the highest international levels of excellence.

www.cambridge.org
Information on this title: www.cambridge.org/9781316513866
DOI: 10.1017/9781108631129

First published 2018

Printed in the United Kingdom by TJ International Ltd. Padstow Cornwall

A catalogue record for this publication is available from the British Library.

ISBN 978-1-316-51386-6 Hardback
ISBN 978-1-108-45670-8 Paperback

Contents

Colour plates can be found between pages 142 and 143.

Contributors

Helen Bailey
Chesapeake Biological Laboratory, University of Maryland Center for Environmental Science, Solomons, MD, USA

Steven J. Bograd
Environmental Research Division, Southwest Fisheries Science Center, National Marine Fisheries Service, National Oceanic and Atmospheric Administration, Monterey, CA, USA

Andreas B. Brink
Joint Research Centre, European Commission (Italy), Ispra, Italy

Graeme M. Buchanan
RSPB Centre for Conservation Science, Royal Society for the Protection of Birds, Edinburgh, UK

Marco Clerici
Joint Research Centre, European Commission (Italy), Ispra, Italy

Monica DeAngelis
National Oceanic and Atmospheric Administration, West Coast Region, Long Beach, CA, USA

Lynn DeWitt
Environmental Research Division, Southwest Fisheries Science Center, National Marine Fisheries Service, National Oceanic and Atmospheric Administration, Monterey, CA, USA

Paula Escribano
Andalusian Centre for the Evaluation and Monitoring of the Global Change (CAESCG), Universidad de Almería, Almería, Spain

Néstor Fernández
German Centre for Integrative Biodiversity Research (iDiv) Halle-Jena-Leipzig, Leipzig, Germany and Estación Biológica de Doñana (EBD-CSIC), Spanish National Research Council, Sevilla, Spain

Karin A. Forney
Marine Mammal and Turtle Division, Southwest Fisheries Science Center, National Marine Fisheries Service, National Oceanic and Atmospheric Administration, Moss Landing, CA, USA

Matthew C. Hansen
University of Maryland Department of Geographical Sciences, College Park, MD, USA

Elliott Hazen
Environmental Research Division, Southwest Fisheries Science Center, National Marine Fisheries Service, National Oceanic and Atmospheric Administration, Monterey, CA, USA and Department of Ecology and Evolutionary Biology, University of California Santa Cruz, Santa Cruz, CA, USA

Mark Hebblewhite
Wildlife Biology Program, Department of Ecosystem and Conservation Sciences, W. A. Franke College of Forestry and Conservation, University of Montana, Missoula, MT, USA

Jennifer Hewson
Betty & Gordon Moore Center for Science, Conservation International, Arlington, VA, USA

Aimee Hoover
Chesapeake Biological Laboratory, University of Maryland Center for Environmental Science, Solomons, MD, USA and National Oceanic and Atmospheric Administration, Pacific Islands Fisheries Science Center, Honolulu, HI, USA

Evan Howell
National Oceanic and Atmospheric Administration, Pacific Islands Fisheries Science Center, Honolulu, HI, USA

Mark Hurley
Idaho Department of Fish and Game, Boise, ID, USA

Ladd Irvine
Marine Mammal Institute, Oregon State University, Hatfield Marine Science Center, Newport, OR, USA

Samuel M. Jantz
University of Maryland Department of Geographical Sciences, College Park, MD, USA

Allison K. Leidner
ASRC Federal, NASA Earth Science Division, Washington DC, USA

Paul Lukacs
Wildlife Biology Program, Department of Ecosystem and Conservation Sciences, W. A. Franke College of Forestry and Conservation, University of Montana, Missoula, MT, USA

Bruce Mate
Marine Mammal Institute, Oregon State University, Hatfield Marine Science Center, Newport, OR, USA

Nicholas J. Murray
School of Biological Sciences, University of Queensland, Queensland, Australia and Centre for Ecosystem Science, University of New South Wales, New South Wales, Australia

Janet Nackoney
University of Maryland Department of Geographical Sciences, College Park, MD, USA

Josh Nowak

Wildlife Biology Program, Department of Ecosystem and Conservation Sciences, W. A. Franke College of Forestry and Conservation, University of Montana, Missoula, MT, USA

Daniel M. Palacios

Marine Mammal Institute, Oregon State University, Hatfield Marine Science Center, Newport, OR, USA

Ilaria Palumbo

Joint Research Centre, European Commission (Italy), Ispra, Italy

Lilian Pintea

The Jane Goodall Institute Department of Conservation Science. Vienna, VA, USA

Antoine Royer

Joint Research Centre, European Commission (Italy), Ispra, Italy

Abdoulkarim Samna

Ministry of Environment and Sustainable Development, Niamy, Niger

Cindy Schmidt

Bay Area Environmental Research Institute, NASA Ames Research Center, Moffett Field, CA, USA

Zoltan Szantoi

Joint Research Centre, European Commission (Italy), Ispra, Italy

Karyn Tabor

Betty & Gordon Moore Center for Science, Conservation International, Arlington, VA, USA

Jessica Wingfield

Chesapeake Biological Laboratory, University of Maryland Center for Environmental Science, Solomons, MD, USA

Preface

The view of Earth from space has provided us with powerful imagery, from the inspiring 'blue marble' image taken by the astronauts of Apollo 17 to disquieting images of progressive deforestation. Images of night-time lights illustrate better than words how much of the terrestrial surface is influenced by humans, while lights from fishing fleets convey the human impact on the ocean. Animations of these features provide a clear visualisation of how rapidly Earth is changing. Beyond the visual impact, the images of the globe that have been collected over decades have scientific applications that enable researchers to examine environmental patterns and processes on land, in the oceans, and in the atmosphere. Through imagery, we can map in space and over time how land cover or sea conditions are changing. This in turn allows us to assess changes in habitats, and determine how the species that rely on these habitats are being impacted. Importantly, research and monitoring enables us to understand and quantify the impact of conservation and restoration actions, allowing modification of activities as needed. We can evaluate if certain actions succeed or fail, in turn creating opportunities for applying such knowledge to other conservation challenges. Images captured from space are making a major contribution to biodiversity conservation. And it comes at a time when we need it more than ever, as Earth's biodiversity is under huge pressure from a multitude of threats.

Conservation problems abound but so do solutions, and satellite remote sensing is contributing to conservation success. A revolution in the number and accessibility of satellite remote sensing observations, combined with advances in computing power and the knowledge that has been gained over the past two decades, has led to increased use and recognition of the technology. There are many examples of how remote sensing data collected from space have helped inform conservation actions, leading to improved outcomes for species or ecosystems. However, the use of satellite remote sensing in conservation is still not generally considered routine or embedded in decision-making protocols. The application of remote sensing to identify conservation problems and solutions, a growing field, aptly named 'conservation remote sensing', is not as straightforward as it sounds. As is the case of any interdisciplinary field, it requires those with different expertise, academic training, and professional backgrounds to learn about and work together on a new topic. For example, conservation scientists themselves have diverse backgrounds that span the dictionary from anthropology through to zoology. They work in distinct ecological and cultural settings, and have had varying training in physical sciences and statistics. Few conservationists have taken remote sensing classes, let alone have an academic background in the field. As a consequence, they are often unfamiliar with key concepts in remote sensing such as trade-offs in spatial, spectral and temporal resolution, or basics such as how satellites 'see' the Earth. Those with a satellite remote sensing background have the knowledge base to exploit satellite instruments with different spatial, spectral, and temporal resolutions, but may have a limited understanding of conservation issues, or what parameters might be of greatest biological importance.

Yet while remote sensing has been applied to conservation research for decades, operational systems and tools for practitioners are still relatively uncommon. This book contains six case studies from around the world, detailing the use of satellite remote sensing for conservation action, followed by a study of the evolution of the use of remotely sensed observations by a conservation non-governmental organisation. Together, they provide examples of how satellite remote sensing has been successfully incorporated into operational conservation. They by no means provide a comprehensive review of studies that are using satellite remote sensing for conservation, but they are all linked in that they demonstrate how Earth observations (a phrase used synonymously with satellite remote sensing) are being used on an ongoing, operational basis. Consequently, they represent a leap from many previous instances where remote sensing played a key, although one-off, role in informing management. Furthermore, the case studies identify how people with diverse backgrounds worked together to advance conservation. Crucially, each project explains how the conditions to bring the groups together came about, discusses the challenges associated with their endeavour, and identifies lessons learned. By providing this developmental insight, we hope to catalyse further uses of remote sensing for conservation.

Our objective is to stimulate collaborations that result in the development and implementation of operational systems through which satellite remote sensing informs conservation. We hope this book will reach both the conservation and remote sensing communities. Both editors come from the ecological side of conservation science. They have worked in conservation research, non-governmental organisations, and space agencies with around 35 years' combined experience in conservation science. Allison started her conservation career studying butterflies in the Rocky Mountains, and only came to remote sensing through an AAAS Science & Technology Policy Fellowship, where she worked at NASA Headquarters in the Earth Science Division. In working with the biodiversity program, she was exposed to the amazing ways that remote sensing advanced biodiversity research and how these observations could be used for conservation applications. Staying on as a contractor after her 2-year fellowship, she increasingly saw the gap between the conservation and remote sensing communities. Graeme started off researching birds in the Scottish uplands and, after time in the Seychelles, Mauritius, and Poland, he returned to his native Edinburgh and upland birds, where he began exploring the potential of satellite remote sensing for mapping upland vegetation. As with Allison, Graeme was exposed to the potential remote sensing in conservation action. His research with The Royal Society for the

Protection of Birds soon took an international direction. We met through a series of workshops aimed to bridge gaps between the conservation and remote sensing communities. We were then inspired to run a session at the 2015 International Congress for Conservation Biology on examples of where satellite remote sensing was being applied to conservation, and were subsequently asked by Cambridge University Press to turn the symposium into a book. As we first set about developing the book, we sharpened our focus to case studies where satellite remote sensing was being used for operational conservation. We also concentrated on conveying the story of how a project came about and lessons learned, in addition to presenting the scientific details.

We developed this book with the goal of reaching those in the conservation and remote sensing communities who are not already working at the boundary of conservation and remote sensing. The conservation community will be familiar with many of the issues presented in each of the six case-study chapters, but may have had less exposure to applications of remote sensing. These same stories will also be valuable to those who have a good practical or academic background in remote sensing, but have less familiarity with how this technology contributes to conservation. Recognising that each of the communities may have variable backgrounds in the other, we included brief primers on conservation and remote sensing. Chapter 1 starts with a short background on biodiversity conservation. Those familiar with conservation need not dwell on the first part of the chapter, but we then provide an overview of conservation remote sensing, followed by an outline of the contributed chapters. For those who need a remote sensing overview or refresher, Brink *et al.*, in Chapter 2, highlight key concepts of satellite remote sensing as it applies to conservation. After the six case-study chapters and a chapter written from the perspective of those working at a conservation non-governmental organisation, we conclude with a chapter that briefly synthesises themes from the preceding chapters and identifies next steps for the field.

We certainly recognise that satellite remote sensing is not a panacea for conservation monitoring, but do firmly believe that there are many more opportunities to apply space-based observations to conservation on the land and in the sea. We hope that the promotion of successful examples, together with the lessons learnt by these studies, will spur new applications of satellite remote sensing to conservation.

Acknowledgements

It goes without saying that we are enormously grateful to all of the chapter authors, as well as the extended conservation and remote sensing communities that enabled their work. The authors invested a great deal of time and energy in writing and editing, and their effort shows. Many also acted as internal peer-reviewers who provided insightful and useful feedback on other chapters. Additionally, we thank Alison Beresford and Amanda Whitehurst, who provided valuable edits, help, and feedback throughout this process and Ilaria Palumbo for early discussions on the content of the book. We thank Cindy Schmidt for help in the final stages of writing. Numerous colleagues in the broader community also provided inspiration and helped us find suitable case studies for this book. We are also grateful to Cambridge University Press, especially Dominic Lewis, for providing the opportunity to share these case studies, and Jenny van der Meijden for help and advice in the closing stages.

Graeme Buchanan – In addition, I thank Paul Donald and Juliet Vickery at The Royal Society for the Protection of Birds for help, opportunities, and general motivation. Pippa Thomson is thanked for her patience. I am indebted to my good friend and co-editor Allison for her management of this book, and for her remaining my friend (I think). I dedicate this book to the memory of friends and colleagues Richard Evans and Tim Cleeves.

Allison Leidner – My colleagues at NASA headquarters provided invaluable support and insights as I was developing, writing, and editing this book. Universities Space Research Association and Arctic Slope Regional Corporation, my former and current direct employer, also assisted. My husband, Andrew Stanton, provided incredible encouragement and help throughout this process, especially with the assistance of B and K. Caitlin Gille also helped with editing. I too am indebted to my good friend and co-editor Graeme for his enduring patience and positive outlook. I dedicate this book to my mentors over the years, who inspired me to pursue a career in conservation and remote sensing: Carol Boggs, Nick Haddad, Jack Kaye, and Woody Turner.

1 A Brief Introduction to Conservation and Conservation Remote Sensing

Graeme M. Buchanan and Allison K. Leidner

1.1 THE CONSERVATION CRISIS

Biological diversity, or biodiversity, is a broad term applied to describe nature. It is formally defined as 'the variability among living organisms from all sources including, inter alia, terrestrial, marine and other aquatic ecosystems and the ecological complexes of which they are part; this includes diversity within species, between species and of ecosystems' by the Convention on Biological Diversity (CBD; CBD 2010). This definition captures the complexity of biodiversity, in that it can represent all levels of biological organisation, from genetic variation to organisms and species, up through communities, ecosystems, and biomes.

Earth's biodiversity is facing a well-documented extinction crisis (e.g. Ceballos *et al.* 2015). The rate of extinction is beyond any natural or 'background' levels, defined as the rate that species had gone extinct over history prior to humans (Pimm *et al.* 1995). As the human population increases, human activities affect ever more places on Earth in ever more diverse ways. Human migration has been documented as having a negative impact on biodiversity for millennia. The settlement of Australia some 50,000 years ago potentially resulted in a rapid depletion of the megafauna of that continent around 45,000 years ago (Van Der Kaars *et al.* 2017). This pattern was also reflected in the extinction of moas (endemic flightless birds) in New Zealand and elephant birds (another group of endemic flightless birds) in Madagascar following the initial settlement of these islands by humans (e.g. Diamond 2013).

1

Numerous studies have documented elevated modern extinction rates in a variety of taxa and geographies, and all point toward the same trend. The current rate of extinction of species is estimated to be two to three orders of magnitude above the background rate (De Vos *et al.* 2015). Out of approximately 21,000 species of birds, mammals, and amphibians extant in 1904, 215 were extinct 100 years later (Baillie *et al.* 2004). Coining the term 'Anthropocene defaunation', Dirzo *et al.* (2014) estimated that 322 species of terrestrial vertebrates have gone extinct since 1500. Extinction rates could be worse for invertebrates, as the same authors estimated a 45 per cent decline in 67 per cent of monitored populations. Notably, these figures are estimates; we cannot know the real figures because we still do not know how many species there are on the planet and do not have anything resembling a species inventory for Earth. New extant species of birds and mammals, perhaps the two best-studied taxa, are still being found. One thing we do know, though, is that plant and animal populations are dwindling. The Living Planet Index, a metric derived from studies of animal populations across the globe, continues to decline (WWF 2016), indicating that extinctions are continuing. The Red List Index, a metric of the number of species facing extinction, continues to deteriorate (IUCN 2017). Understanding why species are declining and going extinct is essential if we are to halt, and then reverse, these declines.

The extinction risk to species can be measured by the International Union for the Conservation of Nature (IUCN) Red List (IUCN 2001). This globally accepted assessment of extinction risk involves the assimilation of information on the various threats that a species is known to face, or is thought to face, based upon expert evaluation. Conclusions on the status of a species are drawn from changes in range size, population sizes, or models quantifying risks of extinction (e.g. population viability analysis). Based upon the worst-case scenario from these criteria, species are assigned a category relating to imminence of extinction, from Least Concern, where the species might still be declining but not a pressing concern, through Near

Threatened, Vulnerable (facing a high risk of extinction in the wild), Endangered (facing a very high risk of extinction in the wild), or Critically Endangered (facing an extremely high risk of extinction in the wild). Fewer than 5 per cent of species have been evaluated with the IUCN criteria but, in 2017, some 24,431 species were identified as Threatened (IUCN 2017). However, even species that are of Least Concern (the lowest-threat category) can experience population or range declines, which can lead to local extinctions and the loss of ecosystem functioning (Ceballos *et al.* 2017).

Habitat alteration and loss are the most prevalent pressures faced by plants and animals (Maxwell *et al.* 2016). The IUCN Red List data indicate that of 8,688 species that are Threatened or Near Threatened, more than 80 per cent are at risk from overexploitation of their populations or the habitat they rely upon. Almost half are at risk from logging of forests while more than half are at risk from agricultural expansion. These figures are of course not mutually exclusive and many species are at risk from both threats, as forests are cleared for agriculture. The massive threat to habitats indicated by these figures in turn highlights the need to have a global assessment of ecosystems status. The IUCN has recently produced a set of criteria for just this purpose (Rodríguez *et al.* 2011, Keith *et al.* 2013). The Red List of Ecosystems will identify the areas on the planet that are suffering from dangerous levels of degradation, which could ultimately lead to ecosystem collapse. The goal is to have a complete assessment for all ecosystems by 2025. Given the high number of species which are of conservation concern, it is perhaps not a surprise that the majority of ecosystems assessed to date have also been identified as being of conservation concern. Indeed, one ecosystem, the Aral Sea, has been so badly degraded that it is now classified as Collapsed, the most severe status that can be assigned to an ecosystem (IUCN-CEM 2016).

Human-induced climate change is a relatively new anthropogenic threat to species and ecosystems. Although the climate of the Earth has changed in the past, it is the rapidity of the current changes that poses a novel threat to biodiversity. There is not enough time for species to adapt by evolving in a given location to cope with changes in temperature and precipitation regimes or to migrate to locations with more suitable climate. One estimate has indicated that between 24 and 50 per cent of bird species, 22 and 44 per cent of amphibians, and 15 and 32 per cent of corals are at risk of extinction due to climate change (Foden *et al.* 2013). Although there is regional variation in the magnitude of climate change, and some species and ecosystems are more vulnerable to such changes than others, few places on land or in the water have been untouched by this force.

To further compound the problem, climate change interacts, often synergistically, with other threats to species and ecosystems (Staudt *et al.* 2013). The deleterious effects of climate change on biodiversity are anticipated to increase in the coming decades, as humans continue to alter the climate (IPCC 2017). Addressing this major threat to ecosystems will remain at the forefront of environmental policy, but 'traditional' threats to biodiversity, such as habitat loss, habitat fragmentation, invasive species, and overhunting and overfishing, are still the main drivers of species endangerment. These threats will likely continue to drive the decline and extinction of species and the degradation of ecosystems.

Extinction of individual plants and animals should be a concern to us all, as extinctions do not occur in isolation from other events. Some species are indicators of the state of an ecosystem, and problems and threats to an entire ecosystem can be manifest first in the decline and loss of such indicator species (e.g. Lindenmayer *et al.* 2000). Some species are keystone species (Paine 1969), upon which many other species within the ecosystem depend, or which play a major role in shaping the ecology of the system. Loss of these species can have cascading implications for other species. Earth's biodiversity, including the ecosystems in which plants and animals live and interact, is the life support system for Earth, and thus for people. Degradation of ecosystems and resultant species loss has consequences beyond the aesthetic and cultural value that some attach to species and wild places. Natural habitats provide ecosystem services that humans rely upon, many of which are expensive to replicate (Costanza *et al.* 1997). These services range from global to local and include climate regulation (forests, soils, and the ocean are sinks for vast quantities of carbon, which, if released, might increase the rate of climate change), water regulation (including clean drinking water and flood management), pollination of crops, and food production (fishing and bushmeat are critical sources of protein for many communities) (Costanza *et al.* 1997).

1.2 CONSERVATION SOLUTIONS

The pressures upon biodiversity are well known and the need for action is recognised. The terms 'conservation', 'conservation biology', and 'conservation science' are used as a 'catch all' to capture the multitudinous actions that are undertaken to understand, maintain, and restore Earth's biodiversity. People have been managing resources in sustainable ways for millennia. Consequently, pinning down the start and foundations of what qualifies as the 'modern' conservation movement is a challenge (Meine 2010). This endeavour is even more daunting if we attempt to summarise this history in a way that is not biased towards a European and North American perspective. Such an undertaking, though incredibly important, would be a book unto itself.

We can look back and see the dates at which various conservation organisations were founded as an indicator of conservation activity. Civil society organisations that promote conservation-related topics have been in existence in one form or another from just before the turn of the twentieth century. In the UK, the organisation that was to become the Royal Society for the Protection of Birds was a fledgling in 1889, and, by the 1920s, the International Council for Bird Preservation, the predecessor of BirdLife International, had been founded. BirdLife International is now a partnership formed of over 100 national biodiversity conservation non-governmental organisations. In the United States, the Wildlife Conservation Society was founded in 1895 and The Nature Conservancy in 1951. Both were preceded, in 1892, by the Sierra Club. Scottish-born John Muir was behind this club, and was also the man who promoted the establishment of Yosemite National Park (California, USA) in 1890, which in turn led to the foundation of the US National Park Service in 1916. However, such a catalogue of dates in which conservation organisations were founded is by no means a satisfactory way to describe the development of the conservation process. It fails to capture actions by governments or by individuals. Neither does it capture rapid expansions or leaps in awareness of the need for conservation. For example, the publication of *Silent Spring* by Rachel Carson in 1962 (by Houghton Mifflin) is often seen as a catalyst for the environmental movement in the USA. We refer the reader to Sodhi and Ehrlich (2010), a free online textbook, for a more thorough history of conservation.

Civil society organisations are not the only groups engaged in conservation. Governmental and non-governmental conservation organisations frequently work together, in collaboration with other sectors of society, to deliver effective conservation. For example, the IUCN (originally called the International Union for the Protection of Nature) was founded in 1948, and its membership is comprised of both governmental and non-governmental organisations, with a focus on providing information and tools to members that enable conservation. The IUCN's efforts led to the Convention on International Trade in Endangered Species of Wild Fauna and Flora (CITES), a species-focused agreement that was adopted by 80 countries and went into force in 1975. That same year, an ecosystem-focused international agreement, the (Ramsar) Convention on Wetlands, also went into effect. The 1992 Rio Earth Summit was a landmark event, from which the CBD was opened for signature. This agreement, signed by 168 nations, calls for conservation of biological diversity, sustainable use of biodiversity, and a fair and equitable sharing of genetic resources derived from biodiversity. In 2010, the CBD strategic plan (CBD 2010) set 20 conservation objectives (named the Aichi targets) to be achieved by signatories by 2020. These address topics including the underlying causes and pressures that result in biodiversity loss, as well as calls to improve the status of biodiversity and enhance its benefits to society.

Conservation involves all levels of society. Together, irrespective of whether they are running a local project or working on pangovernmental environmental conventions, conservationists are working to halt and reverse biodiversity loss and ecosystem degradation. Regardless of the spatial or societal scale at which they are working, the financial and technical resources available to them are often limited. A recent estimate indicated that the funds available to conservationists to meet just two of the 20 Aichi targets (targets 11 and 12) were an order of magnitude lower than what was required (McCarthy *et al.* 2012). Time is also often in short supply. Waiting years to respond to changes in plant or animal populations will only make the problem harder or even impossible to solve. Consequently, the limited resources available to the conservation community need to be targeted to where they are needed most urgently; actions should be based on sound scientific evidence. The field of science that delivers the required evidence can be variously referred to as conservation biology, conservation science, or conservation evidence. There is no one definition of conservation science but Soulé (1985) suggested that 'Conservation biology, a new stage in the application of science to conservation problems, addresses the biology of species, communities, and ecosystems that are perturbed, either directly or indirectly, by human activities or other agents.'

As with biodiversity conservation itself, it is not possible to identify when conservation science emerged as a rigorous, professional discipline. One key professional society, the Society for Conservation Biology, was formed in 1985, and it publishes the journal *Conservation Biology*. The journal *Biological Conservation* began publishing in 1968. However, many of the preceding ecology professional societies and journals made strong contributions to conservation. The British Ecological Society, which was formed in 1913 and is the world's oldest ecological science society, first published the *Journal of Applied Ecology* in 1964.

Conservation does not occur in a vacuum. It requires an understanding of the causes of human activities that lead to conservation problems, and the motivation that people have to see ecosystem services preserved or restored. Conservation science is thus not a strictly biological science. To name just a subset of disciplines, it incorporates anthropology, ecology, natural resource management, psychology, economics, and public policy. The processing and analysis of conservation data has seen computer scientists, mathematicians, and physicists contribute to the field. As you will see in the case studies presented in this book, successful conservation draws upon many disciplines.

Many in the conservation community utilise an adaptive management framework to implement conservation activities (Figure 1.1). This framework integrates research into the design, management, and monitoring of conservation projects in a way that facilitates iterative improvements and adaptation to better achieve the desired outcomes (Salafsky *et al.* 2008). Key steps in this process include identifying a management objective, designing a model of the system, developing a management and monitoring plan to test assumptions, implementing the plans, analysing the results, and then incorporating the findings to optimise future implementation and monitoring (Salafsky *et al.* 2008). Ideally, through routine evaluation and improvement, conservation objectives can be achieved. In reality, the full adaptive management cycle is often not comprehensively implemented, and different aspects of it, such as research and management implementation, are not as integrated as they could be. Satellite remote sensing can potentially play a role in all stages of adaptive management, as evidenced by the case studies presented in subsequent chapters.

1.3   OBSERVING THE EARTH

The first iconic images of the Earth from space were captured during the active days of the 1960s and 1970s. The 1968 Earthrise photograph taken by Apollo 8 astronauts and the 1972 blue marble image (Figure 1.2) were used by the growing environmental movement to

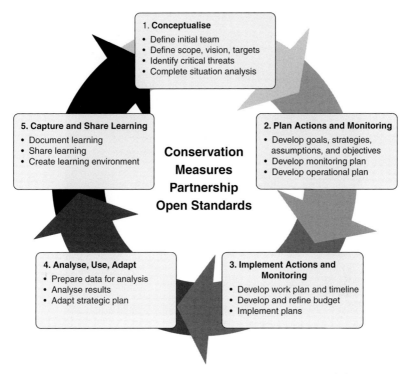

FIGURE 1.1 An example of the adaptive management cycle for conservation.
(Credit Creative Commons.)

rally support. By this point, regular monitoring of the Earth using satellites was already underway. The first non-military satellites were weather satellites, which captured very large-scale images of the Earth on a daily basis. In 1972, the first satellite designed to look at the surface of the Earth, rather than its atmosphere, was launched. The Earth Resources Technology Satellite 1 was the first mission of the pioneering Landsat programme and the ecological community quickly picked up on these Earth observations and other opportunities to explore the processes and functioning of land and sea. However, the cost of images and the need for special computer programmes and computer hardware limited the utilisation of remote sensing images by ecologists and conservationists during this time. High-end

FIGURE 1.2 An image of the Earth, as seen by the Apollo 17 crew.
(Image credit: NASA.) (A black and white version of this figure will appear
in some formats. For the colour version, please refer to the plate section.)

expensive computers were needed to store and process the satellite
images. Often, highly trained specialists were also required to manipu-
late and process the observational data, making them not particularly
conducive to use by conservationists. Despite these significant challenges, some in the ecology and conservation communities realised
early on that satellite remote sensing captured global, repeated data,
which provided a unique perspective on land cover and ocean conditions (Roughgarden *et al.* 1991) and were able to garner the resources to
overcome these challenges. Those who are new to the field of satellite
remote sensing, or people with some knowledge but who want a more
detailed synopsis of the topic, will benefit from reading Chapter 2,
where Brink *et al.* provide an excellent introduction to remote sensing
and the history of its use in conservation. In Chapter 9, Tabor and
Hewson describe how Conservation International, a conservation non-
governmental organisation, has utilised these data over the years.

As remote sensing applications developed in disciplines closely allied to the needs of the conservation community, such as forestry and agriculture, the conservation and remote sensing communities became more aware of each other. Some of the early examples of the application of remote sensing to applied ecology were rooted in agriculture. For example, Wallin *et al.* (1992) used Advanced Very High Resolution Radiometer (AVHRR) observations to identify areas where bird population sizes could explode, and thus where they could become agricultural pests. These collaborations were generally developed on an ad hoc basis, dependent on chance meetings or personal connections. However, the last 15 years have seen an expansion in the field of conservation remote sensing, with growing understanding of the ways in which remote sensing can help inform conservation (Turner *et al.* 2003, Rose *et al.* 2015). One indication of the increased use of remote sensing data by conservation scientists can be seen in the steady increase in the number of scientific papers on conservation remote sensing over this time. In August 2017, a Web of Science search for '"conservation OR biodiversity" AND "remote sensing"' showed a doubling in the number of papers in the last five years (Figure 1.3). We recognise that a search based simply on the term remote sensing will include many other methods for collecting data without visiting a location (e.g. acoustic recording devices or aerial photographs). But a search based upon Landsat, one of the key satellites used by the conservation community, displays a very similar pattern. The increase in the importance of satellite remote sensing in conservation science can be seen in the results of another Web of Science search in August 2017. At this time, about 1.5 per cent of all of the papers returned on a search with 'conservation OR biodiversity' in the subject also had 'remote sensing' in the subject (Figure 1.4). This is over three times the ratio 20 years previously.

Two key review papers published in 2003 (Kerr *et al.* 2003; Turner *et al.* 2003) highlighted ongoing and potential applications of satellite remote sensing to biodiversity and conservation. Turner *et al.* (2003) drew attention to the range of possible applications of remote

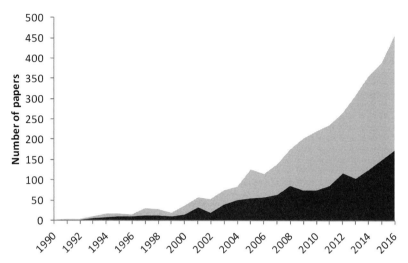

FIGURE 1.3 Number of peer-reviewed journal papers with the subject remote sensing and conservation or remote sensing and biodiversity (light grey) and Landsat and conservation or Landsat and biodiversity (dark grey).

sensing data to conservation monitoring. Importantly, these papers were published in *Trends in Ecology & Evolution*, a journal with a wide readership in the biodiversity, ecology, and evolution research community, as well as the conservation applications community. Consequently, they brought the attention of remote sensing to a new audience.

The transition from scientific research that describes data analysis and potential applications to actual applications and operational decision support is a challenge for any discipline, and conservation is no exception. A plethora of excellent and important literature outlines the impacts of human activities on biodiversity, highlights key areas for land and aquatic protection, and discusses the way in which research has the potential to impact natural resource management. Nevertheless, there is a feeling that 'only lip service is being paid to engagement with end-users in too many cases' (Gibbons *et al.* 2011, p. 507). This is also true for conservation remote sensing. It has remained difficult to get the

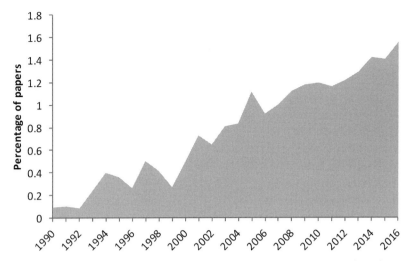

FIGURE I.4 Percentage of peer-reviewed journal papers with the subject conservation or biodiversity that contain the subject remote sensing.

conservation and remote sensing communities to move together and collaborate so as to deliver science that will maximise benefits for conservation. For example, the need and potential for remote sensing to deliver an operational system for monitoring land cover change within important conservation areas or species' ranges was discussed by members of both the conservation and remote sensing communities and was reported in Buchanan *et al.* (2009) but, 6 years later, conservationists were still requesting similar tools (Rose *et al.* 2015).

Rose *et al.* (2015) surveyed conservationists to identify 10 conservation topics that could be informed by the use of satellite remote sensing data (Box 1.1). These topics and the explanation presented by Rose *et al.* highlight how remote sensing can serve conservation beyond being 'just' a monitoring tool. Importantly, the potential applications and uses of remote sensing data represented by these topics have been identified multiple times by different user communities. The results were echoed by three contemporary initiatives. In the UK, conservation practitioners identified roughly similar applications of

---

BOX I.I   **Ten ways remote sensing can contribute to conservation (Rose *et al.* 2015)**

Species distributions and abundances
Species movements and life stages
Ecosystem processes
Climate change
Rapid response
Protected areas
Ecosystem services
Conservation effectiveness
Agricultural and aquacultural expansion and changes in land-use and land cover
Degradation and disturbance regimes

---

remote sensing. Discussions among a global cohort of conservationists, organised in Italy (2013), delivered similar results, as did discussions at a conservation science conference in New Zealand (2011). Combining these initiatives and discussions, Buchanan *et al.* (2015) identified one important addition that fell outside the scope of Rose *et al.*, and this was the need for capacity-building. The conservation community are often unfamiliar with remote sensing, and do not know about the full range of data and potential applications. They do not necessarily know how to use these data, or have access to the appropriate tools to get the best from them (Chapter 3, Chapter 9). This lack of capacity and training is a serious issue (Wegmann 2017). The need is perhaps greatest in the biodiversity-rich tropics, something which is, encouragingly, being addressed to some degree (e.g. De Klerk and Buchanan 2017). Training materials are being made available to the wider remote sensing community, but these are not always delivered to where they are most needed.

These issues notwithstanding, conservation remote sensing has now grown into a discipline in its own right. From an academic perspective, the growth has been so substantial that it now has a dedicated journal as of 2015, *Remote Sensing in Ecology and*

Conservation (Pettorelli *et al.* 2017). The discipline of conservation remote sensing connects a physical science perspective of how to detect information remotely with those seeking to understand how and why biodiversity is changing and what strategies may ameliorate the effects of human pressures on the environment. As with most new areas of study, growth was slow. Initially, remote sensing was often seen as a way to capture an image of a protected area or provide a crude land-cover map, on which ground- or sea-based data on species location were overlaid. These were often one-off applications, on small areas or single time points.

One major barrier to the use of satellite remote sensing for conservation was the cost and accessibility of remote sensing data. Some data, such as the low-resolution AVHRR and Satellite Pour l' Observation de la Terre (SPOT) Vegetation data were, and remain, free. These data were widely used and, as seen in the chapters in this book, remain staples of many conservation remote sensing applications. But it was the spatial resolution of Landsat sensors (30-m pixels) that made it of particular interest and relevance to conservationists. However, until 2008, users had to pay for each Landsat image that they used. The costs were hundreds of dollars (US) per scene (approximately a 185-km by 170-km area; Wulder *et al.* 2008). The high cost of a scene limited the use of remote sensing by non-governmental conservation organisations, both large and small. For many, it was out of reach for one-off assessments, let alone repeated time-series analysis. The decision to make Landsat images free removed a major barrier to the use of remote sensing by the resource-poor conservation community (see Chapter 9 for a more extensive discussion). The value of this decision, and a reaffirmation by space agencies such as National Aeronautics and Space Administration (NASA) to support free and open data policies is praised as a major advance in the use of remote sensing for conservation (Turner *et al.* 2015) and helped set a precedent. The European Space Agency (ESA) has made data from its Copernicus programme free too.

A second major barrier to the use of satellite remote sensing was the storage and processing of very large datasets. Recent advances in computing mean that it has become easier and faster to process the increased volume of data available to users. The advent of cloud computing has seen a further leap in processing speed and power, and an increase in the ability of organisations to store data, or just access them easily online. Software too has advanced. Initially, the image analysis software that is needed to process satellite remote sensing data was prohibitively expensive for conservation organisations. However, free, open-source software packages that can do much of the processing have recently become available. For example, R (www.r-project.org/) and Q-GIS (www.qgis.org/) have provided environments within which data can be processed by individuals with some skill and knowledge. The combination of free software and dedicated training material with worked examples (e.g. Wegmann et al. 2016) should go some way to resolving the current lack of capacity in many areas.

The repeated, global nature of satellite observations, coupled with increased processing capabilities, opens the door to greater possibilities for conservation applications. Continued advances in computing power will see this situation improve. The impact of advances in image availability and computing power are nicely illustrated by an example of the impact of forest loss on biodiversity. In 2008, conservationists from multiple organisations used a limited release of free Landsat data to map forest loss on New Britain (an island in the Bismarck Archipelago, east of New Guinea) and assess tree loss in the ranges of 37 forest birds (Buchanan et al. 2008). Five years later, Hansen et al. (2013) utilised the free access to Landsat images to produce a global map of loss between 2000 and 2012. This enabled some of the authors of the New Britain study to assess tree loss within the ranges of over 10,000 species of amphibians, birds, and mammals across the entire globe, updating the IUCN Red List status of some 800 species (Tracewski et al. 2016). In 8 years they had progressed from a one-off case study on a single island to processing the entire globe, using a method that is readily repeatable to ensure the approach is used in an ongoing, operational way.

The pixel size of freely available satellite data (typically from government organisations) and their derived products (see Chapter 2 for a detailed explanation) ranges from thousands of metres down to tens of metres, with data in the region of 1,000-m to 250-m being those which are collected most frequently. This contrasts with the situation in the field where observers often collect biodiversity data in plot sizes that are centimetres to metres. These observations collected *in situ* can be identified to the species level and life stage, and information can be collected about the size, shape, and condition of an individual organism. This information is invaluable, but monitoring plots or transects on this spatial and taxonomic scale on a regular basis can be challenging and costly, especially in remote areas or in ecological systems that are sensitive to disturbance. Conservation practitioners, for example a park manager in Gabon or a community-based conservation organisation in Indonesia, are often interested to know: 'What is the ecological condition of the area of interest?'; 'How is it changing and why?'; 'How many of a species of concern are there, and is their population size increasing, stable, or decreasing?' Some perceive that remote sensing data cannot contribute to addressing these types of questions because direct species-occurrence measurements cannot typically be made at this resolution. Even areas of conservation concern that are hundreds of square kilometres may be thought to be too 'small' for satellite remote sensing. But although satellites can rarely see individual animals and plants, they can measure characteristics of land or sea surface, such as vegetation structure, photosynthetic activity, or vegetation productivity. Interpretation of these characteristics can be linked to the state, or health, of the ecosystem. If the ecosystem is degraded, then species are potentially at risk. Thus, remote sensing can act as a surrogate where field surveys are not practicable, but it can also complement data collected in the field. For example, the use of remote sensing data in species distribution models has proved to be of great value to conservationists. By linking characteristics measured by satellite remote sensing to the presence or absence of a focal species, or a community of organisms, conservationists can get an insight into their current distribution, identify where a key habitat is being lost, and even find new populations of species.

Conservationists are increasingly interested in larger spatial scales, as the pressures outside of an area of interest greatly influence a park or protected area. Furthermore, the frequency of satellite observations and low cost (though not necessarily easy to access, see above) open up novel opportunities. New remote sensing instruments are increasingly able to collect finer-scale spatial data with a greater diversity of measurement types (Chapter 2) that are relevant for conservationists. This meeting of scales, as well as an increasing awareness of the advantages of satellite remote sensing, have thus reduced the tension between those working with *in situ* and space-based observations, as highlighted in the case studies of this book.

Airborne and drone remote sensing can be used in the collection of higher-resolution spatial data at a scale aligned with field surveys and at times when such information is needed. Unfortunately, airborne remote sensing can be costly, with the price of plane rental, pilot time, and fuel. Unless there is a large budget, airborne surveys are often best at establishing baseline information, which can then be used in conjunction with coarser-spatial-resolution, but higher-temporal-resolution, satellite imagery. Drones are becoming increasingly popular (Koh and Wich 2012) and are also becoming easier and cheaper to use. However, they typically collect limited spectral information, so the information derived from such observations is mostly used to visually detect change on the ground (e.g. logging, poaching, changes in water levels) and spectrally derived vegetation indices. As a consequence, while both technologies are important for conservation, we choose not to focus on them in this book but encourage the reader to explore these observational platforms elsewhere.

1.4 CASE STUDIES HIGHLIGHTING REMOTE SENSING FOR CONSERVATION

Remote sensing provides a growing opportunity to address the expanding and increasingly urgent needs of conservation. But the matching of opportunities to achieve the full potential of conservation remote sensing and desires to push the boundary even further requires the rapid development of productive and sustainable collaborations. The need for, and value of, dialogue between the remote sensing community and the conservation community has been highlighted multiple times (e.g. Buchanan *et al.* 2015, Rose *et al.* 2015). There are now numerous examples of successful collaborations between the two communities. However, the documenting of these examples in the scientific literature necessarily focuses on the technical details of projects and emphasises the scientific concepts that inform conservation actions. These details are incredibly important, but they do not provide the whole story. Journal articles rarely describe the origins of a project, highlight how different types of experts came together to address a problem, analyse the lessons learned, or provide personal perspectives on a project. In order to maximise learning from examples of successful collaborations, we need to communicate not just technical details, but the full 'stories' about projects.

With the case studies in this book, we endeavour to present the full breadth of the collaboration needed to use remote sensing for conservation. Thus, we have asked each team to present not only the scientific and technical aspect of their projects, but also to provide the history of how a collaboration between conservationists and remote sensing scientists came about, the challenges faced and lessons learned, and recommendations for others. Additionally, we sought case studies that highlight the use of remote sensing in an operational way to inform conservation management and action. These examples emphasise what it takes to develop ongoing collaborations, tools, and products. These often require different relationships and discussions from those surrounding one-off remote sensing applications, such as a prioritisation exercise. We hope that sharing knowledge will inspire others to initiate these types of productive collaborations. Finally, while remote sensing plays an important role in conservation-relevant activities such as carbon sequestration (e.g. Reducing Emissions from Deforestation and Degradation; REDD) and law enforcement (e.g. illegal logging, fishing, and mining), we selected case studies that had a more direct biodiversity focus.

At the beginning of this chapter, we provided a brief primer on biodiversity decline and associated conservation efforts, to introduce those on the remote sensing side to the context for the case studies. This introduction was followed by a short history of remote sensing and its applications to conservation, for both conservationists and remote sensing scientists. For those who have a little background in remote sensing, or need a refresher, Chapter 2 by Brink *et al.* gives a very useful overview of the topic, with a focus on applications to conservation. The chapter gives some essential details that will help readers to understand how satellites make observations. Only with this knowledge can the potential (and limitations) of satellite remote sensing be understood. This chapter also directs the reader to additional tools and resources to build greater knowledge on the topic.

The subsequent six chapters are case studies from terrestrial and aquatic systems and form the core of this book. They detail how applications for satellite remote sensing have developed, including the steps taken to ensure that the resultant tools developed are used by the conservation community. The first two chapters focus on the application of satellite remote sensing for monitoring habitats. As satellite remote sensing cannot measure all characteristics to which species are responding, these chapters focus on land cover. In Chapter 3, Murray analyses mudflats in Asia, which are essential habitats for migrating shorebirds that breed in the Arctic. Populations of these birds have declined recently, and these declines have been linked to loss of mudflats. Murray documents how the analysis undertaken by himself and colleagues has contributed to global conservation efforts. In Chapter 4, Jantz *et al.* describe how chimpanzee conservation has benefited from annual assessments of forest-cover change in central Africa. Chimpanzees are threatened with extinction, with habitat loss being a major concern. The appropriate conservation action needs to be targeted to the right location, and two tools developed by Jantz and colleagues have helped inform conservation action for the species. The mapping of forest change was only possible following the free release of Landsat images and advances in cloud computing.

Next are two chapters that consider the application of remote sensing data to sites of conservation interest. In Chapter 5, Palumbo *et al.* describe the long history of using satellite remote sensing to monitor fires across the globe before describing how effective data distribution and easy-to-use tools have helped park managers in Africa to monitor fires. This information is valuable to guide conservation efforts in African protected areas. Active fire data have been available for over a decade, but making these data available in a useful format has been a major advance to conservation. Then, in Chapter 6, Escribano and Fernández outline a remote sensing-based observatory used to guide the management of a protected area in Spain. The Doñana National Park is an internationally known wetland, but threats from surrounding land-use, and activities within the park, create problems for its management. Escribano and Fernández describe how they utilise Moderate Resolution Imaging Spectroradiometer (MODIS) data to describe ecosystem conditions in the park using simple visualisations that distil complex ecosystem function measurements. The data they produce help park managers ensure that the site continues to maintain and restore its unique biodiversity.

The last two case-study chapters detail the use of satellite remote sensing for species management. In Chapter 7, Hebblewhite *et al.* describe how remote sensing is being used to set sustainable hunting levels for mule deer in the western USA. The population dynamics of these deer have been well studied, but the gap between when population data are analysed and hunting quotas need to be set have resulted in inefficient management and has consequences for the entire ecosystem. By using MODIS data, state wildlife managers can obtain timely information on vegetation productivity, which can then be linked to mule deer population size, and thus better inform the setting of hunting quotas. For the final case study in Chapter 8, Bailey *et al.* look at a marine example. They describe how a collaboration between scientists and natural resource managers at a government agency developed a platform that ingests satellite remote sensing data to provide near-real-time alerts about the location of whales off the west coast of the United States. These alerts are used to help reduce ship strikes for the endangered blue whale.

Following these case studies, Tabor and Hewson detail the use of remote sensing data in the operations of Conservation International, an international conservation organisation. This non governmental organisation was a rapid adopter of remote sensing data, and the chapter describes past uses and looks to the future. Finally, the concluding chapter synthesises the lessons identified in the preceding chapters and considers how we can maximise the benefits of remote sensing for conservation in the future.

ACKNOWLEDGEMENTS

We are very grateful to Alison Beresford and Amanda Whitehurst for input to this chapter through comments on various earlier versions.

REFERENCES

Baillie, J., Hilton-Taylor, C., and Stuart, S. N. (2004). *2004 IUCN Red List of Threatened Species: A Global Species Assessment*. Gland: IUCN.

Buchanan, G. M., Brink, A. B., Leidner, A. K., Rose, R., and Wegmann, M. (2015). Advancing terrestrial conservation through remote sensing. *Ecological Informatics*, **30**, 318–321.

Buchanan, G. M., Butchart, S. H., Dutson, G., *et al.* (2008). Using remote sensing to inform conservation status assessment: estimates of recent deforestation rates on New Britain and the impacts upon endemic birds. *Biological Conservation*, **141**, 56–66.

Buchanan, G. M., Nelson, A., Mayaux, P., Hartley, A., and Donald, P. F. (2009). Delivering a global, terrestrial, biodiversity observation system through remote sensing. *Conservation Biology*, **23**, 499–502.

CBD (2010). COP Decision X/2. Strategic plan for biodiversity 2011–2020. See www .cbd.int/decision/cop/?id=12268.

Ceballos, G., Ehrlich, P. R., Barnosky, A., *et al.* (2015). Accelerated modern human-induced species losses: entering the sixth mass extinction. *Science Advances* **1**, e1400253

Ceballos, G., Ehrlich, P. R., and Dirzo, R. (2017). Biological annihilation via the ongoing sixth mass extinction signalled by vertebrate population losses and declines. *Proceedings of the National Academy of Sciences*, **114**, E6089–E6096.

Costanza, R., d'Arge, R., De Groot, R., *et al.* (1997). The value of the world's ecosystem services and natural capital. *Nature*, **387**, 253–260.

De Klerk, H. M. and Buchanan, G. (2017). Remote sensing training in African conservation. *Remote Sensing in Ecology and Conservation*, **3**, 7–20.

De Vos, J. M., Joppa, L. N., Gittleman, J. L., Stephens, P. R., and Pimm, S. L. (2015). Estimating the normal background rate of species extinction. *Conservation Biology*, **29**, 452–462.

Diamond, J. (2013). *The Rise and Fall of the Third Chimpanzee*. London: Random House.

Dirzo, R., Young, H. S., Galetti, M., *et al.* (2014). Defaunation in the Anthropocene. *Science*, **345**, 401–406.

Foden, W. B., Butchart, S. H., Stuart, S. N., *et al.* (2013). Identifying the world's most climate change vulnerable species: a systematic trait-based assessment of all birds, amphibians and corals. *PLOS ONE*, **8**, e65427.

Gibbons, D. W., Wilson, J. D., and Green, R. E. (2011). Using conservation science to solve conservation problems. *Journal of Applied Ecology*, **48**, 505–508.

Hansen, M. C., Potapov, P. V., Moore, R., *et al.* (2013). High-resolution global maps of 21st-century forest cover change. *Science*, **342**, 850–853.

IPCC (2107). IPCC Fifth Assessment Report (AR5). See www.ipcc.ch/report/ar5/ wg2/mindex.shtml.

IUCN (2001). *IUCN Red List Categories and Criteria*. Gland: IUCN.

IUCN-CEM (2016). The IUCN Red List of Ecosystems. Version 2016–1. See http:// iucnrle.org

IUCN (2017). Summary statistics. See www.iucnredlist.org/about/summary-statistics.

Keith, D. A., Rodríguez, J. P., Rodríguez-Clark, K. M., *et al.* (2013). Scientific foundations for an IUCN Red List of Ecosystems. *PLOS ONE*, **8**, e62111.

Kerr, J. T. and Ostrovsky, M. (2003). From space to species: ecological applications for remote sensing. *Trends in Ecology & Evolution*, **18**, 299–305.

Koh, L. P. and Wich, S. A. (2012). Dawn of drone ecology: low-cost autonomous aerial vehicles for conservation. *Tropical Conservation Science*, **5**, 121–132.

Lindenmayer, D. B., Margules, C. R., and Botkin, D. B. (2000). Indicators of biodiversity for ecologically sustainable forest management. *Conservation Biology*, **14**, 941–950.

Maxwell, S. L., Fuller, R. A., Brooks, T. M., and Watson, J. E. (2016). Biodiversity: the ravages of guns, nets and bulldozers. *Nature*, **536**, 143–145.

McCarthy, D. P., Donald, P. F., Scharlemann, J. P., *et al.* (2012). Financial costs of meeting global biodiversity conservation targets: current spending and unmet needs. *Science*, **338**, 946–949.

Meine, C. (2010). Conservation biology: past and present. In Sodhi, N. S. and Ehrlich, P. R., eds., *Conservation Biology for All*. Oxford: Oxford University Press, pp. 7–26.

Paine, R. T. (1969). A note on trophic complexity and community stability. *The American Naturalist*, **103**, 91–93.

Pettorelli, N., Nagendra, H., Rocchini, D., *et al.* (2017). *Remote Sensing in Ecology and Conservation*: three years on. *Remote Sensing in Ecology and Conservation*, **3**, 53–56.

Pimm, S. L., Russell, G. J., Gittleman, J. L., and Brooks, T. M. (1995). The future of biodiversity. *Science*, **269**, 347.

Rodríguez, J. P., Rodríguez-Clark, K. M., Baillie, J. E., *et al.* (2011). Establishing IUCN red list criteria for threatened ecosystems. *Conservation Biology*, **25**, 21–29.

Rose, R. A., Byler, D., Eastman, J. R., *et al.* (2015). Ten ways remote sensing can contribute to conservation. *Conservation Biology*, **29**, 350–359.

Roughgarden, J., Running, S. W., and Matson, P. A. (1991). What does remote sensing do for ecology? *Ecology*, **72**, 1918–1922.

Salafsky, N., Salzer, D., Stattersfield, A. J., *et al.* (2008). A standard lexicon for biodiversity conservation: unified classifications of threats and actions. *Conservation Biology*, **22**, 897–911.

Sodhi, N. S. and Ehrlich, P. R., eds. (2010). *Conservation Biology for All*. Oxford: Oxford University Press.

Soulé, M. E. (1985). What is conservation biology? *BioScience*, **35**, 727–734.

Staudt, A., Leidner, A. K., Howard, J., *et al.* (2013). The added complications of climate change: understanding and managing biodiversity and ecosystems. *Frontiers in Ecology and the Environment*, **11**, 494–501.

Tracewski, Ł., Butchart, S. H., Di Marco, M., *et al.* (2016). Toward quantification of the impact of 21st-century deforestation on the extinction risk of terrestrial vertebrates. *Conservation Biology*, **30**, 1070–1079.

Turner, W., Spector, S., Gardiner, N., *et al.* (2003). Remote sensing for biodiversity science and conservation. *Trends in Ecology & Evolution*, **18**, 306–314.

Turner, W., Rondinini, C., Pettorelli, N., *et al.* (2015). Free and open-access satellite data are key to biodiversity conservation. *Biological Conservation*, **182**, 173–176.

Van Der Kaars, S., Miller, G. H., Turney, C. S., *et al.* (2017). Humans rather than climate the primary cause of Pleistocene megafaunal extinction in Australia. *Nature Communications*, **8**, 14142.

Wallin, D. O., Elliott, C. C., Shugart, H. H., Tucker, C. J., and Wilhelmi, F. (1992). Satellite remote sensing of breeding habitat for an African weaver-bird. *Landscape Ecology*, **7**, 87–99.

Wegmann, M. (2017). *Remote Sensing Training in Ecology and Conservation –* challenges and potential. *Remote Sensing in Ecology and Conservation*, **3**, 5–6.

Wegmann, M., Leutner, B., and Dech, S. (2016). *Remote Sensing and GIS for Ecologists: Using Open Source Software*. Exeter: Pelagic Publishing Ltd.

WWF (2016). Living Planet Report 2016. Risk and resilience in a new era. Gland: WWF International.

Wulder, M. A., White, J. C., Goward, S. N., *et al.* (2008). Landsat continuity: issues and opportunities for land cover monitoring. *Remote Sensing of Environment*, **112**, 955–969.

2 Introduction to Remote Sensing for Conservation Practitioners

Andreas B. Brink, Cindy Schmidt,
and Zoltan Szantoi

2.1 INTRODUCTION AND DEFINITION OF REMOTE SENSING

The aim of this chapter is to introduce the reader to satellite remote sensing. It specifically focuses on information useful to those members of the conservation community that have minimal knowledge of Earth observations. It begins with a brief history of the topic and then provides a technical overview of key remote sensing concepts, with the goal of serving as a foundation for the case-study chapters. Remote sensing is the science of obtaining information about objects, areas, or phenomena from a distance, by a device not in direct contact with the object (Lillesand *et al.* 2014). This chapter primarily focuses on one aspect of remote sensing, which is the use of sensors mounted on Earth-orbiting satellites.

2.2 HISTORY AND MILESTONES OF REMOTE SENSING

The earliest example of airborne remote sensing was when Gaspard-Félix Tournachon took the first aerial photograph in 1858 from a hot-air balloon over the French village of Petit-Becetre (Baumann 2014). Subsequent years saw improvements to photographic technology, and in methods and 'platforms' for acquiring photographs of the landscape. Balloons, kites, and even pigeons were used as carriers for aerial cameras. The Swedish inventor, Alfred Nobel, took the first successful aerial photograph from a rocket-mounted camera in 1897 (Skoog 2010).

The use of aerial photography taken from aeroplanes increased during the First World War, with a focus on military applications. Subsequently, airborne cameras were used for civilian land-survey applications such as topographic and urban mapping, and forest surveys. As a result, techniques and methods for calculating accurate measurements from photographs quickly emerged. The first aerial colour photographs were recorded in the 1930s, preceding their widespread use in both military and civilian applications after the Second World War (Baumann 2014). The more recent development of unmanned aircraft systems (UAS), commonly referred to as drones, has been recognised as an inexpensive aerial photography or video recording platform by the conservation and natural resource management communities and has been used to map habitats, identify and count specific animal species, or track poacher activities (Watts *et al.* 2010).

The emergence of new technologies for measuring the Earth's surface led Evelyn L. Pruitt of the US Office of Naval Research to introduce the term 'remote sensing' in 1960. The subsequent decade represented a new era in the field of Earth observation for several reasons: (i) the recorded imagery moved from analogue photography to digital imagery, processed by computers; (ii) new sensors were able to detect information beyond the visible part of the electromagnetic spectrum, including ultraviolet and infrared; and (iii) platforms shifted from aeroplanes to Earth-orbiting satellites, which were able to routinely cover much larger areas (Baumann 2014). It is these spaceborne sensors that are the focus of this book, and hence much of the rest of this chapter.

Although the first applications of Earth-observing satellites were for military purposes, civilian applications were not far behind. In 1960, the Television Infrared Observation Satellite (TIROS)-1 was launched by the United States as the first meteorological satellite. Subsequent programmes included the Nimbus Weather Satellite Program in 1964 and the Meteosat European Space Agency (ESA) weather satellites series, starting in 1977. The primary goal of these weather satellites was to provide a synoptic view of the Earth's cloud cover and weather patterns on a regular basis. These meteorological satellites inspired and motivated the development of satellites for environmental monitoring.

Towards the end of the 1960s, recognising that observing the Earth's surface from space could provide valuable information for natural resource management and agriculture, Dr William Pecora, then Director of the United States Geological Survey (USGS), proposed a civilian satellite-based remote sensing programme to the United States Secretary of the Interior. This resulted in Project EROS: an Earth Resources Observation Satellite program (Anderson 1975). In 1970, the National Aeronautics and Space Administration (NASA) developed Landsat 1, the first civilian satellite to observe the surface of the Earth (Baumann 2009). Landsat 1 (initially called Earth Resources Technology Satellite 1) was successfully launched on 23 July 1972, and operated until 1978. It was equipped with a new type of sensor known as the Multispectral Scanner (MSS) that provided data in digital matrices, enabling substantial advances in image processing. The Landsat programme, now a collaboration between NASA and the USGS, has proven to be a long-term, reliable source of data for environmental monitoring, with more advanced sensors on successive missions. Six additional Landsat satellites have been successfully launched: Landsat 2 (1975), 3 (1978), 4 (1982), 5 (1984), 7 (1999), and 8 (2013), with more planned (USGS 2016; Figure 2.1). Landsat 6 was launched in 1993 but failed to reach orbit and crashed into the Indian Ocean.

The Landsat programme is the first and the longest continuously operating satellite-based remote sensing system for Earth resources monitoring. It paved the way for subsequent developments in and expansions of satellite remote sensing over the next 40 years. Improvements to optical sensors have resulted in more frequent coverage of the Earth in more detail. New technologies and international cooperation increase the types of observations being made and their application for scientific research and societal benefit. Space agencies, through the Committee on Earth Observation Satellites (CEOS, see Box 2.1), now work together to expand the use of satellite remote

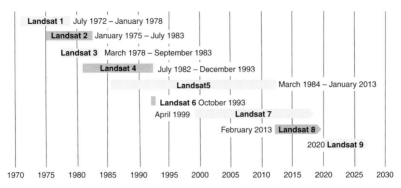

FIGURE 2.1 Operating history of the Landsat programme. (Modified from USGS 2016.)

BOX 2.1   Committee on Earth Observation Satellites (CEOS)

CEOS is the international coordinating body for civilian space agencies. Established in 1984, CEOS promotes cooperation between its 32 space agencies and 28 associate members, which span 22 countries and six continents, with the goal of increasing the utility of space-based observations. Numerous working groups, ad hoc teams, and virtual constellation teams provide information relevant to biodiversity conservation. A relatively new group within CEOS specifically focuses on biodiversity research and conservation applications. This activity is working to build a community that can exploit Earth observations for these purposes, as well as provide feedback on community observing needs to CEOS. More information on CEOS, including the organisation's strategy and work plan, can be found at http://ceos.org/.

---

sensing, leverage historical and ongoing satellite missions, and develop new missions. An in-depth overview and database of all current and future planned Earth observation missions and their instruments are provided by the CEOS Earth Observation Handbook (www .eohandbook.com/). Combined, these instruments provide the ability to better characterise the environment, complementing and extending the Landsat legacy. In the next section, we provide a technical overview of remote sensing and describe key instruments used by the conservation community.

2.3   ELECTROMAGNETIC RADIATION AND THE ELECTROMAGNETIC SPECTRUM

Remote sensing satellites acquire information by detecting electromagnetic energy. In most cases this is energy from the Sun that is reflected, or absorbed then emitted, by objects on the Earth. This energy, called electromagnetic radiation, consists of different wavelengths, which range from high-frequency–short to low-frequency–long wavelengths. The range of the radiation wavelengths (with units measured in metres) is called the electromagnetic spectrum (Figure 2.2). It spans from gamma rays at 10^{-12}-m wavelengths to radio waves at 10^{2}-m wavelengths. Humans can detect only visible light (blue, green, red) with their eyes; however, Earth-observing satellite sensors measure radiation from a much wider range of the electromagnetic spectrum. Most passive sensors (see Section 2.5) measure reflected light in the visible to the infrared regions. The infrared region is divided into distinct wavelength ranges: near-infrared (NIR), mid-infrared (MIR; also called short wave infrared (SWIR)), and thermal infrared (TIR). Active sensors (see Section 2.5) can measure the reflectance of still longer wavelengths of the electromagnetic spectrum. Thus, satellite sensors can 'see' beyond what humans can observe with their eyes.

The Earth's atmosphere influences what kind and how much electromagnetic radiation reaches the Earth's surface. Some wavelengths can pass straight through the atmosphere, while others are absorbed by atmospheric gases such as water, carbon dioxide, oxygen, and ozone. The ability of radiation to pass through the atmosphere is referred to as transmissivity. The parts of the electromagnetic spectrum absorbed by gases in the atmosphere, known as absorption bands, are of limited use for instruments sensing reflected radiation, since little or no radiation from the Sun remains for detection. The opposite of

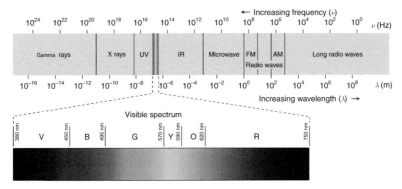

FIGURE 2.2 The electromagnetic spectrum. In the visible region, V = violet, B = blue, G = green, Y = yellow, O = orange, R = red.
(Philip Gringer, Wikimedia Commons.) (A black and white version of this figure will appear in some formats. For the colour version, please refer to the plate section.)

absorption bands are called atmospheric windows. These are the parts of the spectrum where the atmosphere is transparent to specific wavelengths (i.e. there is little or no absorption of radiation by gases). The atmospheric windows are thus the regions of the spectrum where remote sensing instruments typically operate. The atmospheric transmissivity is highest in the visible part of the electromagnetic spectrum as well as parts of the infrared region. Most remote sensing systems are thus designed to measure reflected radiation in the atmospheric windows. However, some sensors, primarily those on meteorological satellites, are designed to measure absorption.

2.4 INTERACTION WITH THE EARTH SURFACE AND THE ATMOSPHERE – SPECTRAL SIGNATURES

Objects on the surface of the Earth reflect, absorb, transmit, or emit electromagnetic radiation, depending on their physical and chemical characteristics. Remote sensing instruments on satellites measure the amount of reflected and emitted radiation from these objects. Reflectance is usually expressed as a percentage of the electromagnetic radiation that is reflected from an object (Lillesand *et al.* 2014). The reflectance of an object is 100% if all radiation striking the object is reflected and 0% if none of the electromagnetic radiation is reflected. The reflectance value of each object usually lies somewhere between 0 and 100% for each wavelength region in the electromagnetic spectrum. Thus, in principle at least, objects on the surface of the Earth can be differentiated from each other by these differences in reflectance. This object-specific reflectance is also defined as the 'spectral signature'. It is analogous to a human fingerprint in that each object has its own spectral signature. However, the ability of a satellite sensor to distinguish these objects depends on its spectral-resolution characteristics, which will be discussed in Section 2.5.

Some objects have very distinct spectral signatures. Healthy green vegetation, for example, absorbs blue and red wavelengths while reflecting green and infrared (especially NIR) wavelengths. Since humans only see reflected radiation in the visible region, we see healthy vegetation as green (Figure 2.3). Water reflects radiation in the shorter blue or green wavelengths in the visible region, which is

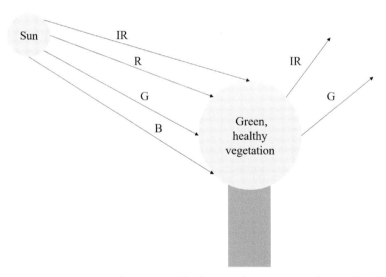

FIGURE 2.3 Absorption and reflection characteristics of green, healthy vegetation. Blue and red radiation is absorbed, while green and infrared radiation is reflected.

the reason water usually appears as blue or blue-green to our eyes; water also absorbs the longer infrared wavelengths. However, water reflectance is also influenced by depth and sediment load, as sediments in the upper layers of the water will result in increased reflectance across all wavelengths. Bare, dry soil has relatively high reflection throughout the visible part of the electromagnetic spectrum compared to other objects. Soils reflect lower in the NIR, but much higher in the MIR than healthy vegetation. Moisture in the soil decreases the reflectance across the wavelengths, but it is particularly sensitive in the MIR wavelengths. Figure 2.4 shows spectral signatures for soil, water, and vegetation as described above. The differences in spectral signatures are used to divide (classify) remotely sensed images into thematic categories or classes. In this way, an image becomes a thematic map (e.g. land-cover map). For vegetation, the ratio between the red (low-reflectance) and NIR (high-reflectance) wavelengths is typically used to calculate a Normalised Difference Vegetation Index (NDVI) to measure vegetation quantity or condition (Rouse *et al.* 1973, Tucker and Sellers 1986). Healthy, green vegetation or areas of high biomass generally have a higher ratio than less healthy vegetation or areas of low biomass. That ratio is usually

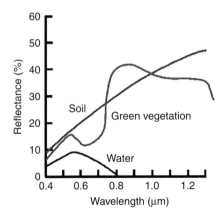

FIGURE 2.4 Spectral signatures of soil, vegetation, and water, and Landsat bands (USGS 2016).

normalised, so the resulting values range from –1 to 1, where –1 to 0 indicates no vegetation and 0 to 1 indicates some quantity of vegetation. The formula for NDVI is $(NIR - R)/(NIR + R)$, where NIR is reflectance in the NIR band and R is the reflectance in the red band.

2.5 SATELLITE AND SENSOR CHARACTERISTICS

It is important to distinguish between sensors (the instruments that make measurements), and satellites (the platforms on which the sensors are mounted). One satellite can have several sensors and different satellites can have the same sensors. For example, the ESA Sentinel-3 satellite has five sensors/instruments including the Ocean and Land Color Instrument (OLCI), the Sea and Land Surface Temperature Radiometer (SLSTR) instrument, a Synthetic Aperture Radar (SAR) altimeter called the SAR Radar ALtimeter (SRAL) instrument, a microwave radiometer (MWR) instrument, and the Doppler Orbitography and Radiopositioning Integrated by Satellite (DORIS) positioning system. All these instruments have different characteristics and purposes.

Orbits

In addition to the sensors, a satellite platform is also described by its orbit. An orbit is the path that a satellite takes around the Earth. The two major orbits are geostationary and polar. Geostationary orbits are approximately 36,000 km above the Earth at the equator. Satellites in geostationary orbits remain in the same place above the Earth as they follow the Earth's rotation. As a result, they continuously observe the same surface section. Geostationary satellites usually have high temporal resolution (in some cases, they observe the same area every 15 minutes) but low spatial resolution (at least 1-km) due to their high altitude. Weather platforms, such as the Meteosat series of satellites, typically have geostationary orbits so they can quickly detect changing weather patterns and monitor severe weather events.

Polar-orbiting satellites are mainly used for environmental monitoring, and have variable spatial resolutions. The term polar orbit describes the satellite path over both polar regions. Many polar-orbiting satellites are Sun-synchronous, which means they consistently pass over a particular location at a specific local time, and therefore are illuminated by the Sun in the same manner. Conditions thus remain constant, ensuring images from different time periods can be easily compared. Polar orbits generally have lower temporal resolutions than the geostationary satellites (in the range of 1 day to 2–3 weeks revisit time), and their orbit altitudes vary between 600 and 900 km. Examples of these satellites include the Landsat satellite, and the Terra and Aqua satellites that carry a Moderate Resolution Imaging Spectroradiometer (MODIS) sensor.

A third type of orbit is the less common near-equatorial, or non-polar, orbit. This type of orbit is typically low-inclination and is used for missions such as the Tropical Rainfall Measuring Mission (TRMM) and the Global Precipitation Measurement (GPM) mission. The low inclination angle (35° for TRMM and 65° for GPM) allows much more frequent observations than polar orbiting satellites, but does not cover the entire Earth surface.

Energy Source

A satellite sensor can either be passive or active. Passive sensors rely on external sources of energy (primarily the Sun) and detect the electromagnetic radiation that is reflected or emitted (Figure 2.5). Conservation practitioners typically rely on passive optical sensors (e.g. sensors on the Landsat and Sentinel-2 satellites). Measurements of reflected electromagnetic radiation are usually only acquired during the day, as they need sunlight to be reflected; however, some sensors can operate at night. For example, the Visible Infrared Imaging Radiometer Suite (VIIRS) instrument day/night band captures observations from reflected moonlight or radiation emitted by anthropogenic lights. Some passive sensors can detect emitted energy from wildfires including the VIIRS thermal band and the Thermal Infrared Sensor (TIRS) on board the Landsat 8 satellite. The longer wavelengths of TIRS energy penetrate smoke so wildfire managers can detect fire perimeters.

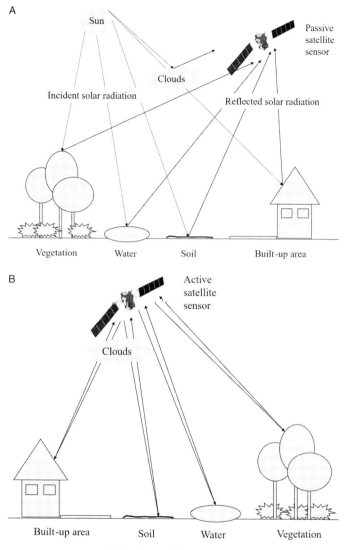

FIGURE 2.5 Passive (A) and active (B) sensor energy and detection flow (note the capability of active (radar) sensors to penetrate clouds).

Active sensors provide their own source of electromagnetic radiation. They send specific wavelengths of radiation to the Earth's surface and then measure the characteristic of the radiation that is reflected back (backscattered radiation) (Graham 1999). Two common active sensors are radar (radio detection and ranging) and lidar (light detection and ranging; Figure 2.5). Radar and lidar sensors work slightly differently. The part of the electromagnetic spectrum utilised by radar sensors lies in the microwave bands (between 1-cm and 1-m). The longer wavelengths mean they have lower energy than the wavelengths measured by passive sensors. Consequently, the minimum spatial resolution of a radar sensor will always be lower than passive sensors that measure higher-frequency wavelengths. Radar sensors generally measure surface roughness, so the rougher the surface, the higher the backscattered light intensity. A flat, smooth surface, such as calm water or a paved road, will have very little backscatter and appears as a dark area in an image. Rough surfaces, such as vegetated areas, produce a lot of backscattered radiation and appear light in an image. Bare soil will vary depending on its surface roughness and moisture content. Radiation from active sensors can penetrate the surface of dry soil or sand, resulting in very little backscatter. Higher moisture content also results in greater backscatter. Since active sensors do not require the Sun for illumination, they can acquire data day or night. They can also penetrate (or 'see through') clouds and smoke. These systems are very useful in areas with persistent cloud cover, such as the tropics. However, the processing of images consisting of backscattered radiation requires specialised software and training. Consequently, radar imagery has been less widely used than images from passive sensors. Examples of radar satellites include Sentinel-1 and Radarsat.

A lidar sensor measures the time it takes for a radiation pulse to hit the ground and return. The sensor can send hundreds of thousands of pulses per second, resulting in a three-dimensional 'point cloud' with millions of points, which can accurately characterise the Earth's surface. One lidar pulse can have many 'returns'. In other words, if a lidar collection occurred in a forested area, the first return from a pulse would be from the top of the canopy, but the pulse would continue down through holes in the canopy and produce multiple returns until it reached the ground, which would be the last return. This not only provides information about bare-ground elevation, but also about vegetation height and structure. Currently, most lidar systems use airborne platforms. NASA's Geoscience Laser Altimeter System (GLAS) instrument, on board the ICESat satellite, was the first satellite lidar instrument. The instrument was primarily used to determine the mass balance of polar ice sheets but has also been used in some vegetation studies (Simard *et al.*, 2011). The sparse spatial coverage of lidar pulses limits the use of GLAS for assessing canopy height or vegetation structure. NASA's Global Ecosystem Dynamics Investigation (GEDI) lidar instrument is due to be launched from Earth in 2019 and will be attached to, and operate from, the International Space Station for 24 months. This instrument will result in approximately 15 billion observations during that time, providing global measurements for forest canopy height, vertical structure, and surface elevation (Dubayah *et al.* 2014).

Spatial, Spectral, Radiometric, and Temporal Resolutions and Their Trade-offs

Remote sensing instrument characteristics can be described by their distinct sensor resolutions, encompassing four types of resolutions: spatial, spectral, radiometric, and temporal (Campbell and Wynne 2011).

Spatial Resolution

The spatial resolution of a sensor can be defined as the separation between two very close features that can be discriminated on the ground, based on picture elements (pixels) that in turn create the two-dimensional array of an image (raster matrix). Hence, pixels are the individual elements of a digital image. They represent an area on the ground known as the resolution cell. More details on the ground can be ascertained with higher spatial resolution. Resolutions vary from low (> 500-m), through medium (500-m–30-m), to high (< 30-m), and to very high (< 5-m). Figure 2.6 shows the same area detected with three different types of sensors.

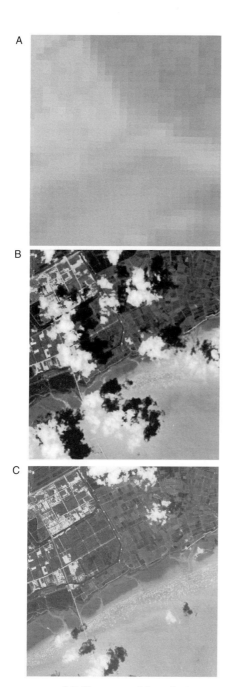

FIGURE 2.6 Different spatial resolutions resulting in different levels of detail with (A) the Terra/Aqua MODIS (300-m), (B) the Landsat 8 OLI (30-m), and (C) the Sentinel-2 MSI (10-m) sensors. Images come from different days in same week. (A black and white version of this figure will appear in some formats. For the colour version, please refer to the plate section.)

There is a direct relationship between image spatial resolution and spatial extent. A higher spatial resolution (i.e. more detailed images) generally implies a smaller spatial extent – swath width – of an image due to the particular sensor recording capabilities. Figure 2.7 shows the swath-size comparison of a low- and a medium-spatial-resolution sensor. Table 2.1 compares the spatial resolutions of different sensors with image swath width. Several high-spatial-resolution images are needed to cover the same area that a single medium-spatial-resolution image can cover. This has cost, computer-processing, and computer-storage implications. Generally, high- and very high-spatial-resolution images are acquired by commercial companies and must be purchased (although Sentinel-2 high-resolution images are free). Many medium- and low-spatial-resolution images are acquired by government organisations, and are generally free to users. Very high- and high-spatial-resolution images have very large file sizes compared to medium- and low-spatial-resolution images. One multispectral (four-band) Worldview tile (Digital Globe, Inc.) with a 1.24-m spatial resolution covers a 64-km^2 area and is 800 MB. By comparison, a multispectral (seven-band) Landsat 8 Operational Land Imager (OLI) image with a 30-m spatial resolution covers a 31,450-km^2 area and is approximately 900 MB.

Spectral Resolution
The spectral resolution of a sensor is the number of spectral bands and their wavelength range (referred to as bandwidth). A band or channel of an image represents the specific range of wavelengths measured by the sensor. The narrower the wavelength range in a given band, the higher the spectral resolution. Images with fewer than 15 bands are generally called multispectral, while images with 15 or more bands, and more typically hundreds of bands, are called hyperspectral. Examples of multispectral imagery include the OLI sensor on Landsat 8, with nine spectral bands, and the Multispectral Instrument (MSI) sensor on board the Sentinel-2 satellite with 13 spectral bands.

Table 2.1 *Comparison of pixel sizes and image swath widths for different sensors*

Organisation, satellite, and sensor	Pixel size (m)	Swath width (km)
NOAA, POES, AVHRR	1,000	2,900
ESA, SPOT, Vegetation	1,000	2,250
ESA, Proba-V, Vegetation	100, 300, 3,000	2,250
EC/ESA, Sentinel-3, OLCI and SLSTR	300 (OLCI), 500–1,000 (SLSTR)	1,270 (OLCI), 1,420 (SLSTR)
NASA, Terra/Aqua, MODIS	250–1,000	2,330
NASA/USGS, Landsat 8, OLI	30 (15 pan)	185
NASA/Japan, Terra, Aster	15 VNIR (30 SWIR)	60
EC/ESA, Sentinel-2, MSI	10, 20, 60	290
ESA, SPOT, Spot-Image	2.5, 5, 10, 20	120
Planet Labs, RapidEye, JSS	5	77
DigitalGlobe, Ikonos, OSA	3.2 (0.82 pan)	11
DigitalGlobe, WorldView-2, WorldView-2	1.84 (0.46 pan)	16.4

Abbreviations: POES, Polar Orbiting Environmental Satellites; AVHRR, Advanced Very High Resolution Radiometer; SPOT, Satellite Pour l' Observation de la Terre. The abbreviation 'pan' indicates panchromatic band resolution (see page 43). Some sensors (e.g. Spot-Image) have different resolutions for different bands.

Hyperspectral sensors – imaging spectrometers – usually have over 200 spectral bands, ranging from visible to SWIR wavelengths. Hyperion, the only spaceborne hyperspectral sensor, was launched on the Earth Observing 1 (EO-1) satellite in 2000, but it was decommissioned in early 2017. EO-1 was a tasking spacecraft, meaning that it only collected data at specific times. Currently, hyperspectral imagery can only be acquired on aircraft platforms. An example of an airborne hyperspectral sensor is the Airborne Visible/Infrared Imaging Spectrometer (AVIRIS). This is primarily used for research projects,

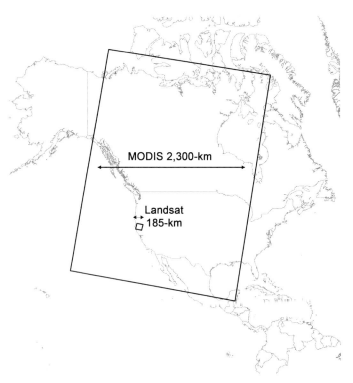

FIGURE 2.7 Footprint of a low-spatial-resolution sensor (MODIS) and a medium-spatial-resolution sensor (OLI).

but the data are publicly available through a data portal (https://aviris.jpl.nasa.gov). Hyperspectral data enable better discrimination between objects on the ground, as they gather more information on their spectral characteristics. They can also identify very important biogeochemical features of plants, such as canopy water content, pigment composition, and pigment content, which provides key information on ecosystem composition and condition (Ustin *et al.* 2004). For most current satellite sensors, there is a trade-off between spatial and spectral resolution. Generally, high-spatial-resolution sensors have four spectral bands that range from the visible to the NIR. Multispectral sensors with moderate spatial resolution, such as OLI on Landsat and the MSI sensor on Sentinel-2, have 10–13 spectral bands that range from the visible to the MIR, and also include TIRS bands. Multispectral sensors often have a panchromatic band, which measures reflectance across a wide range of wavelengths that are otherwise divided across multiple bands. Panchromatic bands tend to have a higher spatial resolution than these other bands. For example, on Landsat 8, the panchromatic band measures reflectance across the blue, green, and red bands (bands 2, 3, and 4), and has a 15-m rather than 30-m resolution.

Figure 2.8 shows the significance of the bandwidths and the number of bands, using a simple example. Water differs from soil and vegetation across all bands. Based only on band 1, soil and green vegetation cannot be distinguished. Band 2 covers a range of wavelengths over which differences between soil and vegetation emerge, but the differences at the longer wavelengths might be lost due to the width of the band. The spectral reflectance values of soil and vegetation can be separated by bands 3 and 4. Thus, a multispectral image, comprising bands 3 and 4 (with possible additional information from band 2), can discriminate between soil and vegetation. However, when discriminating between similar objects, such as vegetation types, finer wavelength ranges and hence finer spectral resolution imagery are required. Hyperspectral

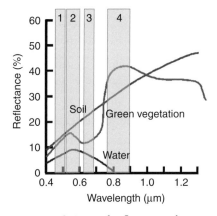

FIGURE 2.8 Spectral reflectance of water, soil, and vegetation and the Landsat 7 Enhanced Thematic Mapper (ETM) bandwidths (only visible and NIR bands 1 to 4 are shown).
(Modified from USGS 2016.).

sensors provide the highest spectral resolution imagery, which detect hundreds of very narrow spectral bands.

Radiometric Resolution

The radiometric resolution of a sensor is defined as the ability of a digital sensor to detect the intensity of reflected electromagnetic radiation (Campbell and Wynne 2011). Intensity values in optical remote sensing terms are the measured solar radiances in a particular band, as reflected from a surface. The sensor can only distinguish between bright and dark (i.e. grey scale). This means that a spectral image is simply a two-dimensional grid, consisting of different grey-scale values. Grey-scale values are measured in bits, where the number of image bits defines the number of grey-scale values a sensor can differentiate. An increased number of bits results in an increased ability to distinguish different features. For example, 1-bit describes a sensor able to differentiate only between black and white, 2-bit has four grey-scale values, while an 8-bit image can have up to 256 values (Figure 2.9 shows increasing levels of grey, by bits). Sensors such as the Landsat family produce at least an 8-bit image. Newer sensors such as the Landsat 8 OLI or the Sentinel-2 MSI record data in the 12-bit range, which can potentially contain up to 4,096 values.

Temporal Resolution

The temporal resolution of a sensor is defined as the time a satellite takes to complete one orbital cycle. Thus, it provides information on the time interval between the acquisitions of two images of the same location on Earth, also known as the revisit time. The higher the temporal resolution, the shorter the revisit time. The revisit time depends on the satellite/sensor capabilities, the swath overlap, and the latitude. Some polar-orbiting satellites have increasing overlap at higher latitudes as orbit paths come closer together near the poles.

Satellite revisit times vary from every few minutes (e.g. geostationary weather satellites), to daily (e.g. the MODIS sensor on the Terra

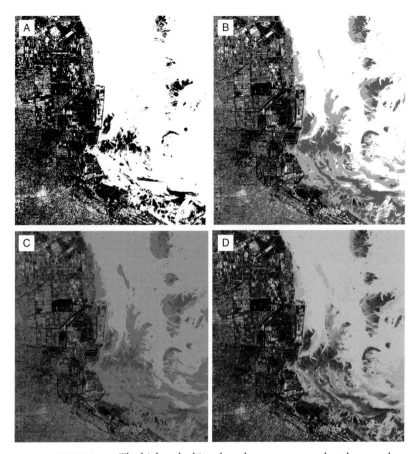

FIGURE 2.9 The higher the bit value, the more grey-scale values can be differentiated by a sensor. (A) 1-bit (two-grey-scale values), (B) 2-bit (four grey-scale values), (C) 3-bit (eight-grey-scale values) and (D) 4-bit (16 grey-scale values).

satellite), to every 16 days (e.g. Landsat 8). Some satellites are able to increase their temporal resolution on request, by tilting and pointing the sensor to a specific area of interest. In these cases, the satellite acquires data of the same area more than one time over an overpass flight or during neighbouring orbit overpasses. Another way of increasing the temporal resolution with moderate- to high-spatial-resolution and low-spatial-extent sensors is to create a multi-satellite constellation system. For example, the Sentinel-2 mission comprises twin polar-orbiting satellites (2A and 2B) in the same orbit, phased at 180° to each other. The Sentinel-2A revisit time is 10 days at the equator; however, including Sentinel-2B (which also has a temporal resolution of 10 days), the combined revisit time is 5 days at the equator, and 2–3 days at mid latitudes.

A single satellite simply cannot make high spatial-, temporal-, and spectral-resolution measurements. Even if it could, the sheer volume of data being sent back to Earth for analysis would become a limiting factor. Consequently, satellite missions and sensors are designed with specific scientific research and/or societal applications in mind. To overcome the limitations of a single sensor, many con-servation practitioners use a multi-sensor approach. For example, NASA's MODIS and ESA's Proba-V sensors, with their high temporal, but low spatial, resolution, can be used to identify NDVI anomalies for a particular year or time of year. A higher-spatial- but lower-temporal resolution sensor such as those on Landsat and Sentinel-2 may then be used to identify the cause of the anomalies in a particular location. Similarly, Landsat or Sentinel sensor imagery may be used to identify areas of land-cover change, but very high-resolution commercial ima-gery can be used to identify the cause of that change.

2.6   DATA PROCESSING LEVELS AND PRODUCTS

Earth observation data acquired by satellite instruments are processed to different levels, prior to distribution to users. These levels range from 0 to 4. Level 0 data are raw and unprocessed. Level 1 data have been geometrically and sometimes radiometrically corrected. Level 2 data are derived geophysical variables and Level 3 and 4 data are derived products. Official definitions for data product levels have been estab-lished by CEOS (CEOS 2008). Level 1 and 2 data products have the highest spatial and temporal resolution, while Level 3 and 4 data are derived products with equal or lower spatial and temporal resolution. From a user perspective, Level 0 ('raw') data are harder to use than higher-level processed data. Some image providers further differentiate the various levels of processing. Sentinel-2 images, for example, are distinguished by Level 1A, 1B, and 1C, although only Level-1C products are distributed to the users. These are radiometrically and geometrically corrected, including being ortho-rectified. This means that the images are corrected for perspective (tilt) and relief (terrain), as well as being spatially registered to a global reference system with sub-pixel accuracy. Level 2A Sentinel-2 images are ortho-images with surface reflectance.

Satellite data need radiometric and atmospheric correction to account for measurement inaccuracies and to improve their ease of use. Sensors record the intensity of radiation from the top of the atmosphere (TOA) as digital numbers. The range of those digital numbers depends on the radiometric resolution of the sensor (Section 2.5). The digital numbers can be converted to radiance using sensor-specific information, which is found in metadata files that are associated with each image. The radiance values can then be converted to TOA reflectance, using information on Earth–Sun distance and solar–zenith angles, which can also be found in the associated metadata files. Reflectance is generally easier to understand because the values range from 0 to 100%. Lastly, since the resulting reflectance values are measured from the TOA, correction for atmospheric conditions is necessary. Removing the effects of the atmosphere results in a surface-reflectance product (Figure 2.10). There are several ways to atmospherically correct an image, including using atmospheric correction models such as COST (Chavez 1996) or dark-object subtraction (Chavez 1988). In the past, users were required to

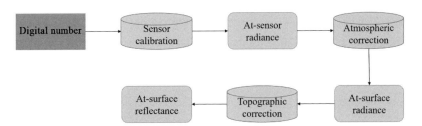

FIGURE 2.10 Radiometric calibration processing steps.

use models or specialised software to conduct their own radiometric and atmospheric corrections, but surface-reflectance products for Landsat 8 and Sentinel-2 data are now available.

Satellite images may also contain geometric distortions that arise from several sources, including orbit errors in the satellite's positioning, the turning of the Earth axis while the satellite is recording, and topographic effects. Such distortions need to be corrected prior to image use. Some distortions are predictable and systematic and can be corrected systematically by the image producer. To improve the geolocation accuracy, ground control points are required to correct and transform the image to a known reference system. Ideally, after geometric correction, the distortion or geometric shift should not exceed the sensor-specific pixel size. Level 1 and higher-level products are radiometrically and geometrically corrected and are made available to the user community. However, for specific applications, advanced users may decide to run their own pre-processing chain to ensure that the image meets their required standards.

2.7   KEY SATELLITE PRODUCTS AND TOOLS FOR BIODIVERSITY CONSERVATION MANAGEMENT AND THEIR APPLICATIONS

Satellite observations have made a substantial contribution to conservation (see Chapter 1). They have helped to identify and monitor areas of high conservation value, to model and predict species distributions, and to characterise community responses to natural and anthropogenic disturbances. The utility of satellite data to the biodiversity and conservation community is directly influenced by several key factors (Leidner *et al.* 2012) including: (i) data continuity (i.e. the long time-series of data); (ii) data affordability (i.e. the free availability of the data), and (iii) data access (i.e. ease of accessing and downloading data). The available expertise to turn the satellite data into useful information is then a final prerequisite to the utility of image data.

Several Level 3 and 4 products derived from satellite images are widely used for biodiversity conservation applications. Examples that are used in later chapters include vegetation indices, snow cover, land surface temperature, and active fire. The NDVI is particularly widely used for assessing vegetation health, for monitoring phenology, or even as an indicator of drought (Pettorelli 2013). As explained in Section 2.4, NDVI uses the differences between reflected NIR and visible red light to measure vegetation 'greenness' and is related to vegetation health and density. The NDVI data from MODIS and, in the near future, VIIRS, are available from web portals such as NASA's Worldview (https://worldview.earthdata.nasa.gov), NDVI data from ESA's Proba-V is available from http://proba-v.vgt.vito.be/en, while NDVI from the Landsat sensors must be generated by individual scientists using geospatial software. Other satellite-derived data, such as sea surface temperature (SST), sea surface height, and chlorophyll-*a*, are critical for studying marine ecosystems.

There are now an increasing number of web-based portals designed for non-specialists, which improve access to satellite-derived products. Notable examples pertinent to subsequent chapters include: (i) Global Forest Watch (GFW, www.globalforestwatch.org), an interactive online forest-monitoring and alert system based on data from multiple sensors (see Chapter 4), which utilises many satellite-derived data (e.g. forest maps by Hansen *et al.* 2013); (ii) Fire Information for Resource Management System (FIRMS) Web Fire Mapper (https://earthdata.nasa .gov/earth-observation-data/near-real-time/firms), which makes use of MODIS and VIIRS data to provide near-real-time data on active fires to natural resource managers (see Chapter 9), and (iii) eStation, which is a service tailored to African countries and that automatically receives, processes, analyses, and distributes remote sensing-derived environmental variables from SPOT Vegetation, Proba-V Vegetation, Meteosat Second Generation Spinning Enhanced Visible and InfraRed Imager (SEVIRI), and Terra/Aqua MODIS sensors (see Chapter 5). Another example is the Global Surface Water Explorer (Pekel *et al.* 2016), an online Landsat-based system that provides global information on the location and temporal distribution of water surfaces over the past 32 years, with visualisation and analysis capabilities.

2.8   SUSTAINABILITY AND LIFESPAN OF REMOTE SENSING SYSTEMS

Continuity, sustainability, and validation of observations are critical for long-term conservation applications. Satellites and their sensors have limited lifespans but, sometimes, satellites will continue to operate well past their design life. For example, Landsat 5 was originally designed to operate for 3 years, but actually operated for a heroic 29 years. However, new observing platforms that continue legacy measurements must be launched, to develop and maintain long-term data records. To ensure consistent measurements between systems and over time, satellite sensors are radiometrically calibrated prior to launch but post-launch calibration is also needed. Satellite data providers assess the quality of data through rigorous and standardised calibration and validation (often abbreviated to cal/val) methods. The radiometric calibration performance of satellite sensors are tested against Earth surfaces with specific characteristics, such as large flat areas of white sands or known dark objects (Teillet *et al.* 2006). CEOS maintains a cal/val portal to ensure standardised methods and measurements across all Earth-observing platforms (http://calvalportal.ceos.org).

Comparison and integration of images from different sensors are often challenging as different instruments have different characteristics. However, cross-calibration of remote sensing sensors is growing in importance due to the increasing need of continuous and long-term environmental monitoring by past, current, and future sensors. The Landsat satellites, for instance, are rigorously cross-calibrated to each other to guarantee that different instruments make comparable measurements for a given pixel. The orbit of Landsat 8 was configured in such a way as to temporarily under-fly Landsat 7 in order to acquire near-simultaneous imaging of the same places. Since Landsat 7 had already been calibrated, the sensors on board Landsat 8 were 'adjusted' to the same values (Roy *et al.* 2014). In the case of Landsat 8, other sensors will be cross-calibrated to its sensors by similar methods. This is particularly pertinent for Sentinel-2, where an accurate cross-calibration between the two systems will enable the development of higher-temporal-resolution products. CEOS currently serves as one venue in which space agencies are coordinating observational continuity. These efforts are critical to ensure consistent, long-term observational records for conservation applications.

2.9 FURTHER READING

This chapter is intended to provide conservationists with a basic introduction to satellite remote sensing, with the goal of helping those new to the topic to better understand the subsequent chapters in this book. In addition to providing a technical foundation, it will also help readers understand the limitations of remote sensing and the trade-offs that users need to consider when exploring the use of satellite data products. This chapter also highlights how advances in remote sensing systems and investment in data products makes Earth observations more accessible to 'new' user communities, such as conservation. Until recently, specialised university courses in remote sensing and expensive computing facilities were a prerequisite to access, process, and analyse satellite images (including often advanced and costly hardware and software). This is no longer the case and we direct those interested in further information to numerous online courses that are available. We list a few key resources after the references in this chapter.

REFERENCES

Anderson, C. A. (1975). *A Biographical Memoir of William Thomas Pecora, 1913–1972*. Washington, DC: National Academy of Sciences.

Baumann, P. R. (2009). History of remote sensing, satellite imagery, part II. See www.oneonta.edu/faculty/baumanpr/geosat2/RS%20History%20II/RS-History-Part-2.html. Accessed 3 January 2017.

Baumann, P. R. (2014). History of remote sensing, aerial photography. See www.oneonta.edu/faculty/baumanpr/geosat2/RS%20History%20I/RS-History-Part-1.htm. Accessed 3 January 2017.

Campbell, J. B. and Wynne, R. H. (2011). *Introduction to Remote Sensing*, 5th edn. New York: Guilford Press.

CEOS (2008). WGISS CEOS Interoperability Handbook. See http://ceos.org/docu ment_management/Working_Groups/WGISS/Documents/WGISS_CEOS-Interoperability-Handbook_Feb2008.pdf. Accessed 26 November 2017.

Chavez, P. S. (1988). An improved dark-object subtraction technique for atmospheric scattering correction of multispectral data. *Remote Sensing of Environment*, **24**, 459–479.

Chavez, P. S. (1996). Image-based atmospheric corrections – revisited and improved. *Photogrammetric Engineering and Remote Sensing*, **62**, 1025–1036.

Dubayah, R., Goetz, S. J., and Blair, J. B. (2014). The Global Ecosystem Dynamics Investigation. American Geophysical Union, Fall Meeting 2014, abstract #U14A-07.

Graham, S. (1999). Remote sensing. NASA Earth Observatory. See http://earthobser vatory.nasa.gov/Features/RemoteSensing/printall.php. Accessed on: 3 June 2017.

Hansen, M. C., Potapov, P. V., Moore, R., *et al.* (2013). High-resolution global maps of 21st-century forest cover change. *Science*, **342**, 850–853.

Leidner, A. K., Turner, W., Pettorelli, N., Leimgruber, P., and Wegmann, M. (2012). Satellite remote sensing for biodiversity research and conservation applications: a Committee on Earth Observation Satellites (CEOS) workshop. See http://remote-sensing-biodiversity.org/images/workshops/ceos/CEOS_SB A_Biodiversity_WorkshopReport_Oct2012_DLR_Munich.pdf.

Lillesand. T., Kiefer, R. W., and Chipman, J. W. (2014). *Remote Sensing and Image Interpretation*. Chichester: Wiley.

Pekel, J. F., Cottam, A., Gorelick, N., and Belward, A. S. (2016). High-resolution mapping of global surface water and its long-term changes. *Nature*, **540**, 418–422.

Pettorelli, N. (2013) *The Normalized Differential Vegetation Index*. Oxford: Oxford University Press.

Rouse, J., Haas R. H., Schell J. A., and Deering, D. W. (1973). Monitoring vegetation systems in the Great Plains with ERTS. In *Third Earth Resources Technology Satellite-1 Symposium*, Washington, DC: National Aeronautics and Space Administration, vol. 1, pp. 309–317.

Roy, D. P., Wulder, M. A., Loveland, T. R., *et al.* (2014). Landsat-8: Science and product vision for terrestrial global change research. *Remote Sensing of Environment*, **145**, 154–172.

Scoog, A. I. (2010). The Alfred Nobel rocket camera. An early aerial photography attempt. *Acta Astronautica*, **66**, 624–635.

Simard, M., Pinto, N., Fisher, J. B., and Baccini, A. (2011). Mapping forest canopy height globally with spaceborne lidar. *Journal of Geophysical Research*, **116**, G04021, 1–12.

Teillet, P. M., Markham, B. L., and Irish, R. R. (2006). Landsat cross-calibration based on near simultaneous imaging of common ground targets. *Remote Sensing of Environment*, **102**, 264–270.

Tucker, C. J. and Sellers, P. J. (1986). Satellite remote sensing of primary production. *International Journal of Remote Sensing* 7, 1395–1416.

USGS (2016). Landsat missions: Imaging the Earth since 1972. See https://landsat .usgs.gov/landsat-missions-timeline. Accessed 16 March 2017.

Ustin, S. L., Roberts, D. A., Gamon, J. A., Asner, G. P., and Green, R.O. (2004). Using imaging spectroscopy to study ecosystem processes and properties. *BioScience*, **54**, 523–534.

Watts, A. C., Perry, J. H., Smith, S. E. *et al.* (2010). Small unmanned aircraft systems for low-altitude aerial surveys. *Journal of Wildlife Management*, **74**, 1614–1619.

WEB RESOURCES

AniMove – Animal Movement and Remote Sensing courses. See http://animove .org/. Accessed 16 March 2017.

Ecosens – Remote Sensing and GIS in Ecology training. See http://ecosens.org/. Accessed 16 March 2017.

ARSET – Applied Remote Sensing Training. See https://arset.gsfc.nasa.gov/. Accessed 16 March 2017

eoPortal Directory. See https://directory.eoportal.org/web/eoportal/satellite-missions;jsessionid=8A5599899960900DB14F206C03786020.jvm1. Accessed 16 March 2017.

ESA, eduspace. See www.esa.int/SPECIALS/Eduspace_EN/index.html. Accessed 16 March 2017.

FIRMS – Fire Information for Resource Management System. See https://earthdata .nasa.gov/earth-observation-data/near-real-time/firms. Accessed 16 March 2017.

eStation – Processing server for the environmental monitoring of land condition in Africa. See http://estation.jrc.ec.europa.eu/. Accessed 16 March 2017.

DOPA – Digital Observatory for Protected Areas. See http://dopa.jrc.ec.europa.eu/ en. Accessed 16 March 2017.

Global Surface Water Explorer. See https://global-surface-water.appspot.com/. Accessed 16 March 2017.

Global Forest Watch (GFW) – www.globalforestwatch.org/about. Accessed 16 March 2017.

3 Satellite Remote Sensing for the Conservation of East Asia's Coastal Wetlands

Nicholas J. Murray

3.1 INTRODUCTION

Tidal flats are typically defined as the area of low-sloping sand, sediment, and muddy deposits that is inundated between low and high tide (Healy *et al.* 2002). They are highly productive coastal wetlands, which provide a broad suite of ecosystem services, including food production, habitat for elements of biodiversity, waste treatment, nutrient cycling, and regulating disturbances from storm events (Levin *et al.* 2001, Syvitski *et al.* 2009). Tidal flats are central to human social values in many coastal communities around the world (Millennium Ecosystem Assessment 2005, Tornqvist and Meffert 2008), partly because they are considered to be one of the world's most productive ecosystem types (Millennium Ecosystem Assessment 2005). With exceptional biodiversity, supported by marine-and terrestrially-derived nutrients, tidal flats host vast reserves of economically important marine organisms (Zedler and Kercher 2005) and support the annual migration of millions of migratory shorebirds around the world (Bamford *et al.* 2008).

Major systems of tidal flats occur adjacent to nearly three-quarters of the world's cities that are considered highly threatened by sea-level rise (Murray *et al.* 2014a). In these areas, they serve a particularly important role; maintenance of their integrity is considered a cost-effective method for protecting coastal communities and infrastructure from sea-level rise and the impacts of climate change (Nicholls *et al.* 2007, Arkema *et al.* 2013, Murray *et al.* 2014a). However, their occurrence in areas that coincide with dense coastal human populations has put them at risk from a variety of human-driven threats. The primary threat to tidal flats over the last century has been coastal development, manifested in the form of large-scale conversion of tidal flats to alternative land-uses such as ports, urban areas, agriculture, and aquaculture (Airoldi and Beck 2007, An et al. 2007, MacKinnon et al. 2012, Murray et al. 2014a). The impacts of so-called 'reclamation' projects can be observed across the world because many major cities, including Auckland, San Francisco, Washington DC, Shanghai, and Copenhagen, were constructed within the intertidal zone (Bird 2010). The reasons for this are unsurprising: (i) tidal flats are cheap to buy and relatively easy to build on, particularly when extending land that has already been developed; (ii) the process of tidal inundation served a purpose to remove waste in cities that did not have well-developed sewage networks; and (iii) their occurrence in the coastal zone has coincided with centres of ocean trade. Even as early as 1900, the great benefits that were achieved with reclamation of coastal wetlands were regularly reported (Beazley 1900).

In addition to direct losses of tidal flats due to reclamation, losses of tidal flats have also been driven by secondary effects of coastal development. Large-scale changes to the coastal environment can severely influence coastal processes (Fagherazzi et al. 2006, Mariotti and Fagherazzi 2013), leading to altered sedimentation regimes (Syvitski et al. 2009, Murray et al. 2014a), and the compaction and subsidence of coastal sediments (Wang 1998, Syvitski et al. 2009, Higgins et al. 2013). Enormous reductions in sediment delivery by the world's major rivers over the past 50 years, mostly caused by river damming (Syvitski et al. 2005, Vorosmarty et al. 2010), has dramatically altered the flux of sediments to the coastal zone. This, in turn, influences the processes of sediment mobilisation and deposition that balance the reductions of tidal flats which are caused by erosion, compaction, subsidence, and sea-level rise (Healy et al. 2002). Advances in the use of satellite-borne radar are beginning to show widespread sinking of the coastal zone, driven by a congruence of threats including compaction, subsidence, and resource extraction (Blum and Roberts 2009, Syvitski et al. 2009). Unfortunately, the ability that tidal flats have to naturally migrate with changing sea levels in some areas has been constrained by extensive coastal development (Kirwan and Megonigal 2013).

3.2   THE TIDAL FLATS OF THE YELLOW SEA, EAST ASIA

Tidal flats are the principal coastal ecosystem that fringes the coastlines of much of Asia (Healy *et al.* 2002), Bird 2010). In East Asia in particular, the widespread loss of tidal flats over the past half century has driven a massive decline in extent, which rivals many of the world's major at-risk ecosystem types. For instance, in the early 1990s, it was estimated that the coastal wetlands of Singapore represented less than 24 per cent of their extent in the 1950s (Hilton and Manning 1995). Over the last 30 years, numerous studies have reported that tidal flats across Asia have diminished in both area and quality (Barter 2002, 2003, An *et al.* 2007, Gilman *et al.* 2008, Kirby *et al.* 2008, Keddy *et al.* 2009, Yang *et al.* 2011). Indeed, a 2012 'situation analysis' commissioned by the International Union for the Conservation of Nature (IUCN) suggested that tidal flats across East and Southeast Asia were a highly threatened ecosystem type, requiring urgent attention by governments, non-governmental organisations (NGOs), and environmental managers to slow their rapid decline (MacKinnon *et al.* 2012).

The Yellow Sea region of East Asia has been documented as a globally important stopover site for migratory birds (Barter 2002, 2003). In the early 1980s, ecologists began to link declines in migratory shorebirds with the loss of tidal flats in China (Close and Newman 1984, Barter 2002, 2003). The tidal flats in the Yellow Sea are among the largest on Earth, extending up to 18-km wide in some areas. They harbour a large number of endemic species, many of which are listed as globally threatened (Murray *et al.* 2015b). The Yellow Sea tidal flats are critical in supporting the annual influx of migratory shorebirds, which complete an annual migration through the 26 countries that make up the East Asian–Australasian flyway. Twice a year, migratory shorebird populations converge in a small number of internationally important shorebird sites in the Yellow Sea. These are the areas that host more than 1 per cent of the global population of a migratory shore-bird species. Crucial to their migration is the thin sliver of the intertidal zone that supports highly productive tidal-flat habitats. These habitats host more than 50 species of shorebirds while they rest and refuel during their long migration from breeding grounds in the Arctic to non-breeding grounds in Asia, Australia, and the Pacific (Barter 2002, Warnock 2010). However, declining numbers of shorebirds migrating through the flyway each year signalled a migration at risk. In the early 2000s, Barter (2003) reported enormous and widespread losses of intertidal habitat within the Yellow Sea's internationally important shorebird staging sites, and echoed earlier suggestions that these losses were the principal cause of shorebird population declines in the East Asian–Australasian flyway. Barter (2003) made a crucial recommendation: that there was an urgent need to better understand the distribution and status of tidal flats across the entire region.

The tidal flats in the Yellow Sea fringe one of the most densely populated coastlines on Earth. Human populations are rapidly migrating to coastal regions of China, North Korea, and South Korea (Seto et al. 2012, Choi 2014, He et al. 2014b), and it has been estimated that more than 160 million people now live in the low-elevation coastal zone (< 10-m above sea level) in the Yellow Sea region alone (Murray et al. 2014a). Such close proximity to rapidly growing human populations has exposed the Yellow Sea tidal flats to considerable anthropogenic pressure over the last few decades. In addition to Barter's warnings in 2003, a number of other studies similarly suggested that tidal flats were rapidly declining in area. For example, Cho and Olsen (2003) reported that more than half of the tidal wetlands that once existed along the coastline of South Korea had been reclaimed for industrial land, port facilities, agriculture, and aquaculture development. A similar picture emerged from China in 2007, which estimated that more than 50 per cent of China's coastal wetlands had been lost since the 1950s, primarily as a result of 'misguided policies' that drove the rapid reclamation of the intertidal zone (An et al. 2007).

Surprisingly, however, there was neither a simple distribution map of the occurrence of tidal flats across the region nor quantitative data that could allow a robust estimate of the amount of tidal flats remaining, in relation to their historical extent. Such data were seen as a high priority to allow the distribution, rates of change, and the conservation status of tidal-flat ecosystems in the Yellow Sea to be quantified (Bamford *et al.* 2008, Murray and Fuller 2012). This lack of data was primarily due to a paucity of methods available for remote sensing the intertidal zone with Earth observation data (Murray *et al.* 2012). Remote sensing methods typically required specific delineation of the waterline at the known time of low tide, enabling intertidal maps to be developed for specific sites or bays (Ryu *et al.* 2002, 2004, Choi *et al.* 2010). Upscaling to country- or continental-scale analyses would have required the acquisition and pre-processing of large numbers of satellite images, necessitating considerable human and computing resources, which were not available at the time. This further limited the application of these methods to anything larger than site scales. Significant new research to overcome these major limitations was clearly required.

Seeing that the lack of quantitative evidence was a major hindrance to the conservation of tidal-flat habitats of the Yellow Sea, as well as to the conservation of migratory shorebirds in the East Asian–Australasian flyway, a small research team from the University of Queensland initiated a remote sensing project to investigate the status of tidal flats in the Yellow Sea region of East Asia. The team comprised ecologists, conservation biologists, mathematicians, remote sensing scientists, and geographers, and their overarching aim was to determine the drivers of migratory shorebird declines in the East Asian–Australasian flyway. In this chapter, I report the role that satellite remote sensing has had in the conservation of the tidal flats in the Yellow Sea. I describe the new remote sensing methods developed for mapping intertidal areas and the discovery of enormous losses of the tidal-flat ecosystem across Yellow Sea. Finally, I analyse the impact that these new data on the status of tidal flats have had on the conservation of the coastal zone and migratory shorebirds.

3.3 TRACKING THE RAPID LOSS OF TIDAL FLATS IN THE YELLOW SEA

The extent of tidal flats changes, over time-frames ranging from minutes (changing tide heights) to years (sediment deposition patterns, coastal processes, and patterns of coastal development). With tidal flats fully exposed only at low tide, most satellite remote sensing efforts have focused on utilising information on the tidal elevation at the time of image acquisition, which in some cases has been coordinated with field work (Ryu *et al.* 2002, 2004, 2008). However, to achieve the aims of this project, it was necessary to develop a new remote sensing method, which could be upscaled efficiently to achieve a consistent time-series of continental-scale maps of the intertidal zone.

Key Definition: The Waterline

Tidal flats are bounded by a waterline, where areas of open water meet exposed tidal flats (Foody *et al.* 2005, Ling *et al.* 2008, Chen and Chang 2009). Precise delineation of the waterline is fundamental for achieving accurate area measurements of tidal flats. However, it is difficult to achieve this, due to the intersection of sediment-laden water with newly exposed benthic substrate. This region of the tidal flat is typically very shallow, subject to wave action, and influenced by local wind conditions, which can confound methods for mapping the instantaneous position of the waterline (Ryu *et al.* 2002). Such methods have included manual interpretation (e.g. manual delineation, manual density slicing) of single image bands or multi-band indices. In the early 2000s, Ryu *et al.* (2002) conducted foundational field work to investigate errors associated with mapping the waterline from satellite data, and showed that 'thresholding' multi-band indices, particularly those that included the near-infrared (NIR) band and a visible band, were the best-performing methods for precisely mapping the position of the waterline. Since then, most methods have followed variations of this approach, whereby an expert-derived pixel value (the threshold) delineates whether pixels are classified as water or land (Ryu *et al.* 2008, Choi *et al.* 2011, Murray *et al.* 2012, 2014a, Ryu *et al.* 2014). Other mapping approaches, such as supervised classifications using tree-based machine-learning methods, object-based remote sensing, and sub-pixel analysis methods, are highly promising for improving the accurate delineation of waterlines in the intertidal zone.

Furthermore, tidal flats consist of a variety of sediment types, ranging from coarse sand to fine muddy sediments (Healy *et al.* 2002). The type of sediment exposed at low tide dictates the amount of water remaining on the tidal flat as it becomes exposed over the tidal cycle (Figure 3.1). Areas of standing water that remain on a tidal flat at low tide are the primary source of omission and commission error in tidal-flat maps. This is because (i) they are mapped as intertidal when they are in fact natural drainage channels, or when the depressions remain inundated throughout the entire tidal cycle; or (ii) they are mapped as water when they are simply wet tidal flats that have not been exposed long enough to adequately dry (Murray *et al.* 2012). High map accuracies are therefore typically achieved with increasing time of exposure (Ryu *et al.* 2002) and on tidal flats with sediment characteristics that enable them to dry rapidly, such as intertidal sand flats (Ryu *et al.* 2004).

Developing a Robust Remote Sensing Method

We considered a range of factors in developing a new remote sensing method for the intertidal zone, including: (i) the cost of data, (ii) achieving high spatial resolution maps, (iii) obtaining adequate data over our region of interest (East Asia), (iv) developing a consistent map time-series over the longest time-frame possible, and (v) the efficiency of implementing the method at continental scales. With these requirements, developing a remote sensing method for use with Landsat Archive data was an obvious choice. Landsat data are free, are of medium (30-m) resolution, and provide the longest space-based

FIGURE 3.1 Reference photographs of tidal flats and typical threats in the Yellow Sea region of East Asia. (A) Tidal flats with extensive dendritic drainage networks with wind chop at the waterline, Gomso Bay, South Korea; note the loss to reclamation for aquaculture. (B) Large, low-sloping tidal flats in Jiangsu Province, China. (C) Coastal reclamation works being implemented in Dandong, Liaoning Province, China. (D) Tidal flats reclaimed for agriculture, urban development, and airport land, Incheon, South Korea.
Adapted from Murray *et al.* 2015a. (A black and white version of this figure will appear in some formats. For the colour version, please refer to the plate section.)

observations of the Earth's surface (Chapter 2). Importantly, the ease of access to Landsat metadata, particularly the time and date of acquisition, enabled the use of regional tide models to estimate the tide height at the time of image acquisition for all images of the region that were available from the Landsat Archive. For each image, we estimated the tide height at image acquisition by running a regional tide model at a central point in the image, and appended the tide height to the image metadata (> 28,000 images between 1972 and 2012). We then classified each image into a two-class land–water image by manually thresholding the normalised difference water index (NDWI).

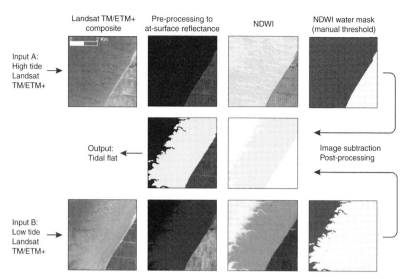

FIGURE 3.2 Remote sensing workflow developed for mapping changes in the extent of East Asia's tidal flats. Landsat Archive imagery is ingested, pre-processed, and classified into a two-class land–water image, which is then differenced among images acquired at high and low tide.
Adapted from Murray *et al.* 2012. (A black and white version of this figure will appear in some formats. For the colour version, please refer to the plate section.)

Next, we selected all high- and low-tide images, which were identified as those that were acquired within the upper and lower 10 per cent of the observed tidal range. Lastly, the extent of the intertidal zone was identified by differencing high- and low-tide image pairs within a 3-year period (Figure 3.2).

The remote sensing method was suitable for use with all Landsat Archive data, and was implemented in the Python programming language, allowing the intertidal zone for each Landsat footprint to be processed by an experienced analyst in ~30 minutes. Scaling the workflow up to the entire coastlines of China, North Korea, and South Korea, nevertheless took several months to account for image downloading (~ 1 month), storage (we used a 10 TB server), post-processing, and mosaicking into a final one-class intertidal product. Significant further time was required for validation against independent data and for the analysis of extent and change, which we completed in Python and R.

It was necessary to make a number of key assumptions in the mapping process. Most importantly, it was unlikely that Landsat scenes were acquired at the precise moment of low tide, so the best cloud-free images within the upper and lower 10 per cent of the tidal range over a 3-year time-frame were used for each time period (Murray *et al.* 2012). This approach tended to result in conservative estimates of area, because the full extent of tidal flats is always likely to be larger than the extent mapped using satellite remote sensing. Despite potentially conservative estimates of area, this remote sensing approach achieves unbiased estimates of tidal-flat extent due to its standardised implementation. This allows it to be used to map tidal flats and estimate change in their extent over any period for which there are Landsat Archive data.

Quantifying Losses of Tidal Flats in the Yellow Sea

The use of the full Landsat Archive allowed the extent of tidal flats to be mapped for the first time (Figure 3.3), allowing area changes over a 30-year period to be quantified, from the mid 1980s to around 2010 (Murray *et al.* 2014a, 2014b, 2015b). Overall, my colleagues and I discovered vast losses of tidal flats (Figure 3.4); our results suggested that tidal flats had declined in area by 30 per cent over the 4,000-km of coastal study area in just 30 years. Comparisons with historical topographic maps developed in the 1950s from aerial-survey imagery suggested that nearly two-thirds of tidal flats had been lost over the last five decades (Murray *et al.* 2014a). Losses of this magnitude indicate that tidal flats in the Yellow Sea have been declining at a rate that rivals many of the world's most at-risk ecosystems, such as tropical forests, mangroves, and tropical peatlands (Figure 3.4). Indeed, tidal flats in the Yellow Sea declined at a rate of 0.79 per cent per year, faster than the rate of global deforestation in the 1990s (0.52 per cent per year; Achard *et al.* 2002). These results, which represented the first quantitative estimates of the amount of tidal flats remaining in the Yellow

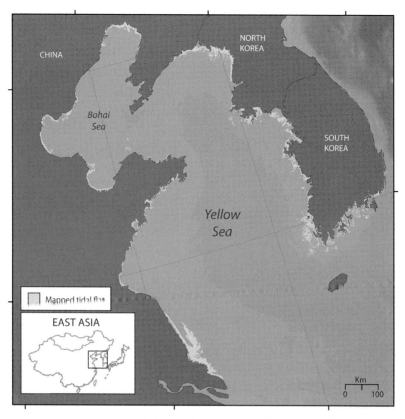

FIGURE 3.3 Map of the distribution of tidal flats in the Yellow Sea, East Asia. The map was developed using the remote sensing method depicted in Figure 3.2.
Adapted from Murray *et al.* 2015b.

Sea, were corroborated by estimates derived from reports published by national governments in the region (Murray *et al.* 2014a). For the first time, high-resolution maps, which were robust, repeatable, and transparent, unequivocally showed that tidal flats in the Yellow Sea had rapidly declined. This would not have been possible without satellite remote sensing data. With the 30-m-resolution maps of the region and an analysis of their losses now complete, we sought to deliver the results to scientists, governments, and NGOs, with the aim of ultimately influencing coastal conservation strategies in the region.

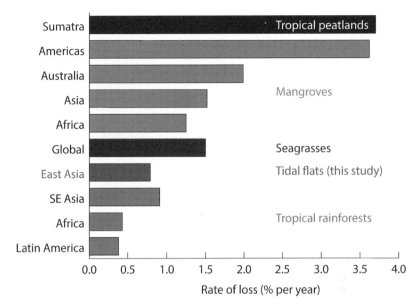

FIGURE 3.4 Comparison of global ecosystem loss rates per region with the decline of tidal flats in mainland East Asia.
Data for other ecosystems were acquired from the published literature (Valiela *et al.* 2001, Achard *et al.* 2002, Waycott *et al.* 2009, Koh *et al.* 2011, Miettinen *et al.* 2012).

3.4 THE IMPACT OF REMOTE SENSING ON CONSERVATION IN THE YELLOW SEA

With the publication of the results of the remote sensing project beginning in 2012, ultimately culminating in the publication of several scientific papers, reports, and datasets (MacKinnon *et al.* 2012, Murray and Fuller 2012, Murray *et al.* 2012, 2014a, 2014b, 2015a, 2015b, Murray and Fuller 2015), a range of conservation actions, aimed at reducing losses of tidal flats in the Yellow Sea region or slowing the decline of migratory birds were initiated (Table 3.1). The singular contribution of our particular remote sensing effort is difficult to quantify, particularly given the vast and ongoing conservation efforts of NGOs, intergovernmental organisations, and individuals across the region. Here, I focus on a few tangible contributions

Table 3.1 *Evidence suggesting that data obtained by remote sensing resulted in conservation action for tidal-flat habitats in Asia and migratory shorebirds. Confidence indicates whether or not the conservation action occurred.*

Action	Evidence	Confidence	Source
Supporting the meetings of parties of the East Asian–Australasian Flyway Partnership (EAAFP)	These data were presented at four EAAFP meetings of partners, which enabled direct communication with decision-makers from 16 countries that are formal partners of the EAAFP	High	(Murray and Fuller 2012)
Listing of eight species of migratory shorebird under Australia's flagship environmental protection legislation, the Environmental Protection and Biodiversity Conservation Act (1999)	Cited as evidence supporting the listings of species under threatened species listing criteria A1 and A2	High	Listing advice for several species (Department of Environment 2014b, 2014a)
Listing of four shorebird species on the Concerted and Cooperative Action List maintained under the Convention for Migratory Species	Cited as supporting evidence when assessing criterion (i) Conservation Priority, and criterion (iv) Confidence in the Science for red knot and bar-tailed godwit	High	(Leyrer *et al.* 2014)
Improved protected-area management in China	Delivery of results demonstrating poor protected-area performance to meetings of provincial environmental managers and decision-makers	Low	(Murray and Fuller 2015)
Listing of the Yellow Sea tidal-flat ecosystem as Endangered on the IUCN Red List of Ecosystems	Supported assessment under Criterion A (reduction in area) of the IUCN Red List of Ecosystems	High	(Murray *et al.* 2014a, 2015a, 2015b)

that may not have been possible without the high-quality information on the distribution, extent, and change of tidal flats on mainland East Asia.

Communication and Outreach

The East Asian–Australasian Flyway Partnership (EAAFP) was instrumental in disseminating the results of the remote sensing analysis to governments and NGOs working along the flyway. Representing a partnership of 35 governments, international NGOs, and intergovernmental organisations, the partnership maintains a list of internationally important wetland sites as part of the EAAFP site network, and serves as a major facilitator of coordinated conservation for migratory shorebirds. A key activity of the EAAFP is convening several technical working groups, which provide a useful pipeline to deliver the results of research projects to the partners themselves. Furthermore, the partnership convenes biennial meetings of partners to report against the EAAFP strategy, to discuss conservation issues and priorities, and to foster collaboration across the flyway. From 2010 to 2017, members of the University of Queensland research team have attended the meetings as scientific experts, which has allowed the identification of research and conservation issues for migratory shorebirds in the East Asian–Australasian flyway, including the problem of ongoing intertidal habitat loss. The dissemination of the results of the remote sensing project have led to the development of new collaborations with researchers across the flyway. In addition to regular delivery of results at the EAAFP meetings of parties, the project results have also been delivered in a range of workshops, formal meetings, and conferences across Asia. This ongoing effort to communicate our research was accompanied by a clear message, comparing the rate of tidal-flat loss in the Yellow Sea with rates of other flagship ecosystems, including tropical forests and mangroves (Figure 3.4). In addition, many of the subsequent analyses of the datasets were focused on highlighting key issues that could trigger conservation action, such as a comparison of the performance of China and South

Korea's coastal protected-area network in conserving the intertidal zone (Murray and Fuller 2015). Our analysis of spatially explicit rates of loss within and outside protected areas indicated that, while protected areas effectively included tidal-flat ecosystems across China and South Korea, habitat losses continued to occur inside some protected areas for the duration of our analysis (~30 years). This was particularly an issue for China, where the rate of habitat loss within protected areas was similar to that outside protected areas. This suggests that protection of habitats within these areas is not being enforced, and that tidal-flat ecosystems are not being preserved. This was not the case in South Korea, where designation was effective at halting the ongoing degradation of tidal flats. With this finding, a clear conservation action was identified and communicated in the workshops with government officials: namely, to improve the management of coastal protected areas in order to ensure that habitat losses within protected areas are immediately halted.

Conservation Contributions

During the remote sensing analysis, a major international report was commissioned by the IUCN. The 'IUCN situation analysis on East and Southeast Asian intertidal habitats, with particular reference to the Yellow Sea (including the Bohai Sea)', was a major synthesis of all knowledge related to the loss of tidal-flat habitat across East and Southeast Asia, and was the first place where the analysis of the remote sensing data was published (MacKinnon *et al.* 2012). The purpose of the situation analysis was to provide context to a motion forwarded at the 2012 IUCN World Conservation Congress held in Jeju, South Korea. The motion itself was aimed at achieving agreement to take the actions needed to conserve migratory shorebirds in the East Asian–Australasian flyway, and included statements relating to the changing extent of intertidal habitats in the Yellow Sea region, including 'NOTING that the rate of loss of intertidal wetlands is particularly severe around the Yellow Sea (as much as 50% in the last 30 years) and is continuing in key areas across the flyway'. The motion was ultimately approved as resolution 5.028: 'Conservation of the East Asian–Australasian flyway and its threatened waterbirds, with particular reference to the Yellow Sea'. It signalled agreement, by more than 1,300 member organisations of the IUCN, that the degradation of tidal-flat habitats in the Yellow Sea was a critical issue for migratory shorebirds. While the ultimate impact of resolution 5.028 is difficult to quantify, it acted as a catalyst to improve the coordination of conservation activities in the flyway by laying the foundation for further workshops, meetings, and reports over the next 5 years. Four years later, a supplementary motion, specifically directed at improving conservation of tidal-flat habitats across the flyway, entitled 'Conservation of intertidal habitats and migratory waterbirds of the East Asian–Australasian flyway, especially the Yellow Sea, in a global context', was agreed at the 2016 IUCN World Conservation Congress in Honolulu, Hawaii.

A second indication of conservation actions supported by the remote sensing analysis was the contribution our results made to the listing of migratory shorebirds in Australia under national threatened species legislation. With rapid and ongoing declines of shorebirds detected across Australia (Clemens *et al.* 2016, Studds *et al.* 2017), many species were assessed for eligibility to be listed on the primary environmental legislation, the Environment Protection and Biodiversity Conservation Act 1999. Like the listing criteria for the IUCN Red List of Threatened Species, Australia's species listing protocol includes criteria for assessing the extent of population declines over a given time-frame. For each of the eight migratory shorebird taxa nominated between 2014 and 2016, a structured assessment of their risk of extinction was made, using all available information. A key piece of evidence used to interpret population declines was the amount of habitat loss across the migratory pathways of the birds. In particular, because a large proportion of the Australian population of many migratory shorebirds converge at staging sites in the Yellow Sea, habitat loss in that region was a strong line of evidence in the listing assessments. For example, losses of intertidal habitat detected within staging sites used by the Far Eastern curlew (*Numenius mada-gascariensis*), which were quantified in this project using Landsat Archive data, supported a listing of Critically Endangered under the population decline criterion. Indeed, the listing of the Far Eastern curlew included an assessment under Criterion A2(a), which states 'Population reduction observed, estimated, inferred or suspected in the past where the causes of the reduction may not have ceased OR may not be understood OR may not be reversible'. The conservation advice indicated that listing under A2 was possible, based on popula-tion declines detected from count data, and supported by remote sensing data that indicated widespread and ongoing losses of intertidal habitat in the Yellow Sea (hence A2(a)). Arising directly from the Critically Endangered listing, and again heavily citing the habitat-loss work, an international single-species action plan for Far Eastern curlew conservation was approved by the 9th Meeting of the Partners of the EAAFP in 2017, which includes state and non-state actors, working across its entire migratory distribution, and aims to 'restore the Far Eastern curlew's population to a positive growth rate for a period of at least three generations'.

Data from the remote sensing project also supported motions to add several species of migratory shorebirds to the Concerted and Cooperative Action List maintained under the Convention for Migratory Species (CMS). This list is intended to provide a rapid mechanism to improve the status of species listed in Appendix 2 of the CMS, which has a purpose to identify threatened migratory spe-cies that require cooperative agreements to ensure their survival (Leyrer *et al.* 2014). The remote sensing findings were cited as evi-dence for two of the listing criteria: (i) Conservation Priority and (ii) Confidence in the Science, which were considered strong with the use of remote sensing data. Again, the impact of this action list is difficult to quantify, but it is likely contributing to conservation prioritisa-tions, an improved coordinated action for the conservation of migra-tory species, and an improved ability to manage coastal ecosystems across Asia in a cooperative way.

Finally, the listing of the Yellow Sea tidal-flat ecosystem on the IUCN Red List of Ecosystems marked the first ecosystem in Asia to be classified according to its risk of collapse under the new global standard for listing ecosystems (Keith *et al.* 2015, Murray *et al.* 2015b, Rodríguez *et al.* 2015, Bland *et al.* 2016). Assessment under Criterion A of the Red List of Ecosystems, which assesses changes in the extent of an ecosystem as a symptom of ecosystem decline, indicated that more than 50 per cent but less than 80 per cent of the ecosystem has been lost over a 50-year period (Criterion A), qualifying it for listing as globally Endangered (Murray *et al.* 2015b). The assessment of Criterion A is most commonly achieved with remote sensing data, which generally give high confidence in the listings (Rodríguez *et al.* 2015). Listings of ecosystems according to their risk of collapse (Least Concern through to Critically Endangered) provide a powerful tool for clear communication of a critical conservation problem, and will enable governments, international NGOs, and intergovernmental partnerships to clearly identify the ecosystems at most need of conservation interventions (Keith *et al.* 2015).

3.5 FUTURE RESEARCH PRIORITIES

Preserving coastal wetlands is crucial for protecting coastlines from several climate-driven threats, including sea-level rise and more intense cyclones (Tornqvist and Meffert 2008, Syvitski *et al.* 2009, Nicholls and Cazenave, 2010). Recently, coastal ecosystems were shown to be both an inexpensive and effective way to shield human communities from storm surges, coastal erosion, severe storms, and other consequences of climate change (Arkema *et al.* 2013). For coastal ecosystems that are under increasing pressure from rapidly growing coastal populations, sea-level rise, and spreading degradation (Michener *et al.* 1997, Alongi 2002, Duarte 2002, Diaz and Rosenberg 2008, Kirwan *et al.* 2010, Pendleton *et al.* 2012), an improved knowledge on the state of coastal ecosystems is vital. With the emergence of the enormous computing power that can be accessed in a cloud-computing environment, remote sensors are now able to utilise and analyse datasets that are far larger than previously possible. For instance, the recent development of global-scale high-resolution forest-change and water-occurrence products required the processing of 650,000 Landsat images, and demonstrated an emerging ability to complete remote sensing studies at scales that were previously impossible (Hansen *et al.* 2013, Pekel *et al.* 2016). The use of these types of geospatial analysis systems for mapping tidal flats is a high priority, because it would solve several issues that limited the size of our analysis. For instance, the ability to utilise all of the images that have ever been acquired in the coastal zone will enable coastlines to be mapped with higher precision and at a range of tide heights (Li and Gong 2016).

The extent of tidal flats have been decreasing as a result of a multitude of threatening processes. The major ones, such as coastal reclamation for agriculture, urban development, aquaculture, and ports, are well known around the world to be a threat to coastal ecosystems (Millennium Ecosystem Assessment 2005, Nicholls *et al.* 2007, MacKinnon *et al.* 2012). However, other threats, including subsidence, compaction, sediment depletion, vegetation loss, and erosion, have all been implicated in tidal-flat losses, suggesting that declines of coastal ecosystems are due to a complex suite of interacting processes. Several mechanistic modelling studies have shown that tidal flats are susceptible to declines in sediment (Fagherazzi *et al.* 2006, Mariotti and Fagherazzi 2013) or vegetation loss (Kirwan and Murray 2007, Kirwan *et al.* 2008, Gedan *et al.* 2011). Other studies have shown erosion of tidal flats (up to 150-m per year) following declines in sediment outflow from rivers (Yang *et al.* 2001, 2003, Wang *et al.* 2012), and major subsidence of coastal areas following resource extraction (Bi *et al.* 2011, Higgins *et al.* 2013). Therefore, further research to identify the specific drivers of tidal-flat loss (beyond coastal reclamation) would significantly contribute to resolving any remaining uncertainty related to the causes of the loss of these ecosystems in East Asia.

The new scientific methods developed in this project have catalysed a range of new studies, which have aimed to map the intertidal habitats in areas where this has not been done before (Dhanjal-Adams et al. 2016), to assess coastal dynamics (Liu et al. 2012, 2013, Li and Gong 2016), to assess protected-area coverage (Dhanjal-Adams et al. 2015), and to monitor land reclamation (He et al. 2014b, Chen et al. 2016). Furthermore, these results have been used to investigate the impact of habitat loss on migratory shorebird populations (Clemens et al. 2016, Murray et al. 2017, Studds et al. 2017).

3.6 CONCLUDING REMARKS

Here, I have shown that a single remote sensing study has tangibly supported conservation of an ecosystem type, has contributed to the conservation of one of the world's least understood coastal ecosystems, and is leading to improved management of important tidal flats in the East Asian–Australasian flyway. However, it is difficult to quantify the exact conservation outcomes that were implemented in response to the newly available high-quality data on the distribution and change of tidal flats across the region. The problem of tracking the conservation impact of scientific studies is widespread in the field of conservation sciences (Cook et al. 2013). To meet the enormous import and export demand that drives their surging economies, the governments of China and South Korea are pursuing rapid coastal development strategies, which are rendering much of their coastlines devoid of natural intertidal habitats. These losses have been driven by (i) reclamation of coastal wetlands for urban, industrial, and agricultural land, (ii) the expansion of activities associated with resource extraction, (iii) the indirect effects of large-scale changes to river catchments, and (iv) ongoing processes such as erosion and subsidence (Murray et al. 2014a). Over the next 15 to 20 years, plans to construct sea walls, ports, aquaculture farms, and industrial areas across the region's coastline will double the amount of habitat already reclaimed by human development (MacKinnon et al. 2012). With coastal urban areas, including some of the largest urban areas in the world, projected to expand by 2030 (Seto *et al.* 2012, He *et al.* 2014a), vast losses of biodiversity and ecosystem services seem inevitable (Cho and Olsen 2003, Zhao *et al.* 2004, An *et al.* 2007, Wang *et al.* 2010). It is therefore critical to continue work to (i) improve methods for monitoring tidal flats, (ii) identify research gaps and conduct research that will better allow the processes driving tidal-flat change to be identified, and (iii) identify and implement conservation actions that will reduce losses of tidal flats throughout the region. In particular, improving the amount of Landsat Archive data being analysed would allow a longer time-frame to be considered, with higher confidence, and would simply require increased computing resources and improved access to pre-processed satellite data. Similarly, reducing the time-lag between the acquisition of satellite data and the dissemination of results to governments, NGOs, and environmental managers would perhaps be the greatest catalyst to improved coastal conservation in Asia. This could be achieved through the development of an automatic monitoring and alert system, such as those available for monitoring deforestation (Martin *et al.* 2012, Hansen *et al.* 2016).

REFERENCES

Achard, F., Eva, H. D., Stibig, H-J., *et al.* (2002). Determination of deforestation rates of the world's humid tropical forests. *Science*, **297**, 999–1002.

Airoldi, L. and Beck, M. W. (2007). Loss, status and trends for coastal marine habitats of Europe. In Gibson, R. N., Atkinson, R. J. A., and Gordon, J. D. M., eds., *Oceanography and Marine Biology, Volume 45*. Boca Raton, FL: Taylor & Francis, ch. 7.

Alongi, D. M. (2002). Present state and future of the world's mangrove forests. *Environmental Conservation*, **29**, 331–349.

An, S. Q., Li, H. B., Guan, B. H., *et al.* (2007). China's natural wetlands: past problems, current status, and future challenges. *Ambio*, **36**, 335–342.

Arkema, K. K., Guannel, G., Verutes, G., *et al.* (2013). Coastal habitats shield people and property from sea-level rise and storms. *Nature Climate Change*, **3**, 913–918.

Bamford, M., Watkins, D., Bancroft, W., Tischler, G., and Wahl, J. (2008). Migratory shorebirds of the East Asian–Australasian flyway: population estimates and internationally important sites. Canberra: Wetlands International – Oceania.

Barter, M. (2003). The Yellow Sea: a race against time. Wader Study Group Bulletin, 100.

Barter, M. A. (2002). Shorebirds of the Yellow Sea: importance, threats and conservation status. Wetlands International Global Series 9, International Wader Studies 12. Canberra: Wetlands International – Oceania.

Beazley, A. (1900). The reclamation of land from tidal waters. *Nature*, **1603**, 266–267.

Bi, X., Wang, B., and Lu, Q. (2011) Fragmentation effects of oil wells and roads on the Yellow River Delta, North China. *Ocean & Coastal Management*, **54**, 256–264.

Bird, E. (2010). *Encyclopedia of the World's Coastal Landforms*. Berlin: Springer.

Bland, L. M., Keith, D. A., Miller, R. M., Murray, N. J., and Rodríguez, J. P. (2016). Guidelines for the application of IUCN Red List of Ecosystems Categories and Criteria, Version 1.0. Gland: International Union for the Conservation of Nature.

Blum, M. D. and Roberts, H. H. (2009). Drowning of the Mississippi Delta due to insufficient sediment supply and global sea-level rise. *Nature Geoscience*, **2**, 488–491.

Chen, W. W. and Chang, H. K. (2009). Estimation of shoreline position and change from satellite images considering tidal variation. *Estuarine Coastal and Shelf Science*, **84**, 54–60.

Chen, Y., Dong, J., Xiao, X., *et al.* (2016). Land claim and loss of tidal flats in the Yangtze Estuary. *Scientific Reports*, **6**, 24018.

Cho, D. O. and Olsen, S. B. (2003). The status and prospects for coastal management in Korea. *Coastal Management*, **31**, 99–119.

Choi, J. K., Ryu, J. H., Lee, Y. K., *et al.* (2010) Quantitative estimation of intertidal sediment characteristics using remote sensing and GIS. *Estuarine Coastal and Shelf Science*, **88**, 125–134.

Choi, J. K., Oh, H. J., Koo, B. J., Ryu, J. H., and Lee, S. (2011). Crustacean habitat potential mapping in a tidal flat using remote sensing and GIS. *Ecological Modelling*, **222**, 1522–1533.

Choi, Y. R. (2014). Modernization, development and underdevelopment: reclamation of Korean tidal flats, 1950s–2000s. *Ocean & Coastal Management*, **102**, 426–436.

Clemens, R. S., Rogers, D. I., Hansen, B. D., *et al.* (2016). Continental-scale decreases in shorebird populations in Australia. *Emu*, **116**, 119–135.

Close, D. and Newman, O. (1984). The decline of the eastern curlew in south-eastern Australia. *Emu*, **84**, 38–40.

Cook, C. N., Mascia, M. B., Schwartz, M. W., Possingham, H. P., and Fuller, R. A. (2013). Achieving conservation science that bridges the knowledge–action boundary. *Conservation Biology*, **27**, 669–678.

Department of Environment (2014a). Consultation document on listing eligibility and conservation actions: *Calidris ferruginea* (curlew sandpiper). Canberra: Department of Environment.

Department of Environment (2014b). Consultation document on listing eligibility and conservation actions: *Numenius madagascariensis* (eastern curlew). Canberra: Department of Environment.

Dhanjal-Adams, K., Hanson, J., Murray, N., *et al.* (2015). Distribution and protection of intertidal habitats in Australia. *Emu*, **116**, 208–214.

Dhanjal-Adams, K. L., Mustin, K., Possingham, H. P., and Fuller, R. A. (2016). Optimizing disturbance management for wildlife protection: the enforcement allocation problem. *Journal of Applied Ecology*, **53**, 1215–1224.

Diaz, R. J. and Rosenberg, R. (2008). Spreading dead zones and consequences for marine ecosystems. *Science*, **321**, 926–929.

Duarte, C. M. (2002). The future of seagrass meadows. *Environmental Conservation*, **29**, 192–206.

Fagherazzi, S., Carniello, L., D'alpaos, L., and Defina, A. (2006). Critical bifurcation of shallow microtidal landforms in tidal flats and salt marshes. *Proceedings of the National Academy of Sciences*, **103**, 8337–8341.

Foody, G. M., Muslim, A. M., and Atkinson, P. M. (2005). Super-resolution mapping of the waterline from remotely sensed data. *International Journal of Remote Sensing*, **26**, 5381–5392.

Gedan, K., Kirwan, M., Wolanski, E., Barbier, E., and Silliman, B. (2011). The present and future role of coastal wetland vegetation in protecting shorelines: answering recent challenges to the paradigm. *Climatic Change*, **106**, 7–29.

Gilman, E. L., Ellison, J., Duke, N. C., and Field, C. (2008). Threats to mangroves from climate change and adaptation options: A review. *Aquatic Botany*, **89**, 237–250.

Hansen, M. C., Potapov, P. V., Moore, R. *et al.* (2013). High-resolution global maps of 21st-century forest cover change. *Science*, **342**, 850–853.

Hansen, M. C., Alexander, K., Alexandra, T. *et al.* (2016). Humid tropical forest disturbance alerts using Landsat data. *Environmental Research Letters*, **11**, 034008.

He, C., Liu, Z., Tian, J., and Ma, Q. (2014a). Urban expansion dynamics and natural habitat loss in China: a multiscale landscape perspective. *Global Change Biology*, **20**, 2886–2902.

He, Q., Bertness, M. D., Bruno, J. F., *et al.* (2014b). Economic development and coastal ecosystem change in China. *Scientific Reports*, **4**, 5995.

Healy, T., Wang, Y., and Healy, J., eds. (2002). *Muddy Coasts of the World: Processes, Deposits, and Function*, Amsterdam: Elsevier Science.

Higgins, S., Overeem, I., Tanaka, A., and Syvitski, J. P. M. (2013). Land subsidence at aquaculture facilities in the Yellow River Delta, China. *Geophysical Research Letters*, **40**, 3898–3902.

Hilton, M. J. and Manning, S. S. (1995). Conversion of coastal habitats in Singapore: indications of unsustainable development. *Environmental Conservation*, **22**, 307–322.

Keddy, P. A., Fraser, L. H., Solomeshch, A. I., *et al.* (2009). Wet and wonderful: the world's largest wetlands are conservation priorities. *BioScience*, **59**, 39–51.

Keith, D. A., Rodríguez, J. P., Brooks, T. M., *et al.* (2015). The IUCN Red List of Ecosystems: motivations, challenges, and applications. *Conservation Letters*, **8**, 214–226.

Kirby, J. S., Stattersfield, A. J., Butchart, S. H. M., *et al.* (2008). Key conservation issues for migratory land- and waterbird species on the world's major flyways. *Bird Conservation International*, **18**, S49–S73.

Kirwan, M. L. and Murray, A. B. (2007). A coupled geomorphic and ecological model of tidal marsh evolution. *Proceedings of the National Academy of Sciences of the United States of America*, **104**, 6118–6122.

Kirwan, M. L. and Megonigal, J. P. (2013). Tidal wetland stability in the face of human impacts and sea-level rise. *Nature*, **504**, 53–60.

Kirwan, M. L, Murray, A. B., and Boyd, W. S. (2008). Temporary vegetation disturbance as an explanation for permanent loss of tidal wetlands. *Geophysical Research Letters*, **35**, L05403.

Kirwan, M. L., Guntenspergen, G. R., D'alpaos, A., *et al.* (2010). Limits on the adaptability of coastal marshes to rising sea level. *Geophysical Research Letters*, **37**, L23401.

Koh, L. P., Miettinen, J., Liew, S. C., and Ghazoul, J. (2011). Remotely sensed evidence of tropical peatland conversion to oil palm. *Proceedings of the National Academy of Sciences*, **108**, 5127–5132.

Levin, L. A., Boesch, D. F., Covich, A., *et al.* (2001). The function of marine critical transition zones and the importance of sediment biodiversity. *Ecosystems*, **4**, 430–451.

Leyrer, J., Van Nieuwenhove, N., Crockford, N., and Delany, S. (2014). Proposal for adding four subspecies of bar-tailed godwit to the CMS Cooperative Action List during the 2014–2017 triennium. Convention on Migratory Species, Quito, Ecuador, 4–9 November.

Li, W. and Gong, P. (2016). Continuous monitoring of coastline dynamics in western Florida with a 30-year time series of Landsat imagery. *Remote Sensing of Environment*, **179**, 196–209.

Ling, F., Xiao, F., Du, Y., Xue, H. P., and Ren, X. Y. (2008). Waterline mapping at the subpixel scale from remote sensing imagery with high-resolution digital elevation models. *International Journal of Remote Sensing*, **29**, 1809–1815.

Liu, Y., Li, M., Cheng, L., Li, F., and Chen, K. (2012). Topographic mapping of offshore sandbank tidal flats using the waterline detection method: a case study on the Dongsha Sandbank of Jiangsu Radial Tidal Sand Ridges, China. *Marine Geodesy*, **35**, 362–378.

Liu, Y., Li, M., Zhou, M., Yang, K., and Mao, L. (2013). Quantitative analysis of the waterline method for topographical mapping of tidal flats: a case study in the Dongsha Sandbank, China. *Remote Sensing*, **5**, 6138–6158.

Mackinnon, J., Verkuil, Y. I., and Murray, N. J. (2012). IUCN situation analysis on East and Southeast Asian intertidal habitats, with particular reference to the Yellow Sea (including the Bohai Sea). *Occasional Paper of the IUCN Species Survival Commission*, No. 47.

Mariott, G. and Fagherazzi, S. (2013). Critical width of tidal flats triggers marsh collapse in the absence of sea-level rise. *Proceedings of the National Academy of Sciences*, **110**, 5353–5356.

Martin, T. G., Nally, S., Burbidge, A. A., et al. (2012). Acting fast helps avoid extinction. *Conservation Letters*, **5**, 274–280.

Michener, W. K., Blood, E. R., Bildstein, K. L., Brinson, M. M., and Gardner, L. R. (1997). Climate change, hurricanes and tropical storms, and rising sea level in coastal wetlands. *Ecological Applications*, **7**, 770–801.

Miettinen, J., Shi, C., and Liew, S. C. (2012). Two decades of destruction in Southeast Asia's peat swamp forests. *Frontiers in Ecology and the Environment*, **10**, 124–128.

Millennium Ecosystem Assessment (2005). *Ecosystems and Human Well-being: Current State and Trends*. Washington, DC: Island Press.

Murray, N. J. and Fuller, R. A. (2012). Coordinated effort to maintain East Asian–Australasian flyway. *Oryx*, **46**, 479–480.

Murray, N. J. and Fuller, R. A. (2015). Protecting stopover habitat for migratory shorebirds in East Asia. *Journal of Ornithology*, **156**, 217–225.

Murray, N. J, Phinn, S. R., Clemens, R. S., Roelfsema, C. M., and Fuller, R. A. (2012). Continental scale mapping of tidal flats across East Asia using the Landsat Archive. *Remote Sensing*, **4**, 3417–3426.

Murray, N. J., Clemens, R. S., Phinn, S. R., Possingham, H. P., and Fuller, R. A. (2014a). Tracking the rapid loss of tidal wetlands in the Yellow Sea. *Frontiers in Ecology and the Environment*, **12**, 267–272.

Murray, N. J., Wingate, V. R., and Fuller, R. A. (2014b). Mapped distribution of tidal flats across China, Manchuria and Korea (1952–1964). *Pangaea*, https://doi.org/10.1594/PANGAEA.837090.

Murray, N. J., Clemens, R. S., Phinn, S. R., Possingham, H. P., and Fuller, R. A. (2015a). Threats to the Yellow Sea's tidal wetlands. *Bulletin of the Ecological Society of America*, **96**, 346–348.

Murray, N. J., Ma, Z., and Fuller, R. A. (2015b). Tidal flats of the Yellow Sea: A review of ecosystem status and anthropogenic threats. *Austral Ecology*, **40**, 472–481.

Murray, N. J., Marra, P. P., Fuller, R. A., *et al.* (2017). The large-scale drivers of population declines in a long-distance migratory shorebird. *Ecography*, doi: 10.1111/ecog.02957.

Nicholls, R. J. and Cazenave, A. (2010). Sea-level rise and its impact on coastal zones. *Science*, **328**, 1517–1520.

Nicholls, R. J., Wong, P. P., Burkett, V. R., *et al.* (2007). Coastal systems and low-lying areas. In Parry, M. L., Canziani, O. F., Palutikof, J. P., Linden, P. J. V. D., and Hanson, C. E., eds., *Climate Change 2007: Impacts, Adaptation and Vulnerability. Contribution of Working Group II to the Fourth Assessment Report of the Intergovernmental Panel on Climate Change.* Cambridge: Cambridge University Press, pp. 315–356.

Pekel, J. F., Cottam, A., Gorelick, N., and Belward, A. S. (2016). High-resolution mapping of global surface water and its long-term changes. *Nature*, **540**, 418–422.

Pendleton, L., Donato, D. C., Murray, B. C., *et al.* (2012). Estimating global "blue carbon" emissions from conversion and degradation of vegetated coastal ecosystems. *PLOS ONE*, 7, e43542.

Rodríguez, J. P., Keith, D. A., Rodríguez-Clark, K. M., *et al.* (2015). A practical guide to the application of the IUCN Red List of Ecosystems criteria. *Philosophical Transactions of the Royal Society B*, **370**, 20140003.

Ryu, J. H., Won, J. S., and Min, K. D. (2002). Waterline extraction from Landsat TM data in a tidal flat: a case study in Gomso Bay, Korea. *Remote Sensing of Environment*, **83**, 442–456.

Ryu, J. H., Na, Y. H., Won, J. S., and Doerffer, R. (2004). A critical grain size for Landsat ETM+ investigations into intertidal sediments: a case study of the Gomso tidal flats, Korea. *Estuarine Coastal and Shelf Science*, **60**, 491–502.

Ryu, J. H., Kim, C. H., Lee, Y. K., *et al.* (2008). Detecting the intertidal morphologic change using satellite data. *Estuarine Coastal and Shelf Science*, **78**, 623–632.

Ryu, J., Nam, J, Park, J,, *et al.* (2014). The Saemangeum tidal flat: long-term environmental and ecological changes in marine benthic flora and fauna in relation to the embankment. *Ocean & Coastal Management*, **102**, 559–571.

Seto, K. C., Güneralp, B., and Hutyra, L. R. (2012). Global forecasts of urban expansion to 2030 and direct impacts on biodiversity and carbon pools. *Proceedings of the National Academy of Sciences*, **109**, 16083–16088.

Studds, C. E., Kendall, B. E., Murray, N. J., *et al.* (2017). Rapid population decline in migratory shorebirds relying on Yellow Sea tidal mudflats as stopover sites. *Nature Communications*, **8**, 14895.

Syvitski, J. P. M., Vörösmarty, C. J., Kettner, A. J., and Green, P. (2005). Impact of humans on the flux of terrestrial sediment to the global coastal ocean. *Science*, **308**, 376–380.

Syvitski, J. P. M., Kettner, A. J., Overeem, I., *et al.* (2009). Sinking deltas due to human activities. *Nature Geoscience*, **2**, 681–686.

Tornqvist, T. E. and Meffert, D. J. (2008). Sustaining coastal urban ecosystems. *Nature Geoscience*, **1**, 805–807.

Valiela, I., Bowen, J. L., and York, J. K. (2001). Mangrove forests: one of the world's threatened major tropical environments. *BioScience*, **51**, 807–815.

Vorosmarty, C. J., Mcintyre, P. B., Gessner, M. O., *et al.* (2010). Global threats to human water security and river biodiversity. *Nature*, **467**, 555–561.

Wang, X. A., Chen, W. Q., Zhang, L. P., Jin, D., and Lu, C. Y. (2010). Estimating the ecosystem service losses from proposed land reclamation projects: a case study in Xiamen. *Ecological Economics*, **69**, 2549–2556.

Wang, Y. (1998). Sea-level changes, human impacts and coastal responses in China. *Journal of Coastal Research*, **14**, 31–36.

Wang, Y. P., Gao, S., Jia, J., *et al.* (2012) Sediment transport over an accretional intertidal flat with influences of reclamation, Jiangsu coast, China. *Marine Geology*, **291–294**, 147–161.

Warnock, N. (2010). Stopping vs. staging: the difference between a hop and a jump. *Journal of Avian Biology*, **41**, 621–626.

Waycott, M., Duarte, C. M., Carruthers, T. J. B. *et al.* (2009). Accelerating loss of seagrasses across the globe threatens coastal ecosystems. *Proceedings of the National Academy of Sciences*, **106**, 12377–12381.

Yang, H-Y., Chen, B., Barter, M., *et al.* (2011). Impacts of tidal land reclamation in Bohai Bay, China: ongoing losses of critical Yellow Sea waterbird staging and wintering sites. *Bird Conservation International*, **21**, 241–259.

Yang, S-L., Ding P-X., and Chen, S-L. (2001). Changes in progradation rate of the tidal flats at the mouth of the Changjiang (Yangtze) River, China. *Geomorphology*, **38**, 167–180.

Yang, S-L., Belkin, I. M., Belkina, A. I., *et al.* (2003) Delta response to decline in sediment supply from the Yangtze River: evidence of the recent four decades and expectations for the next half-century. *Estuarine, Coastal and Shelf Science*, **57**, 689–699.

Zedler, J. B. and Kercher, S. (2005). Wetland resources: status, trends, ecosystem services, and restorability. *Annual Review of Environment and Resources*, **30**, 39–74.

Zhao, B., Kreuter, U., Li, B., *et al.* (2004). An ecosystem service value assessment of land-use change on Chongming Island, China. *Land Use Policy*, **21**, 139–148.

4 Global Forest Maps in Support of Conservation Monitoring

Samuel M. Jantz, Lilian Pintea, Janet Nackoney, and Matthew C. Hansen

4.1 INTRODUCTION

Changes in forest cover affect the delivery of important ecosystem services including climate regulation, carbon storage, water quality and supply, and biodiversity richness (Vitousek 1997, Avissar and Werth 2005, Jantz *et al.* 2014, Pimm *et al.* 2014). More than one-third of global forest cover has been lost (Defries 2012); rates of forest loss in the tropics, Earth's main reservoir for terrestrial biodiversity, are rising (Myers *et al.* 2000, Mittermeier *et al.* 2005; Kim *et al.* 2015). Forest loss, whether from fire, wood harvest, or clear-cutting for agriculture, remains a major threat to species survival (Brooks *et al.* 1999; Brook *et al.* 2006; Hilton-Taylor *et al.* 2008). In several studies, conservation scientists have employed the species area relationship to estimate the number of species that could eventually become extinct because of forest habitat loss (Pimm and Askins 1995, Brooks *et al.* 1999, Strassburg *et al.* 2012). Due to recent advances in technology and computing, we can now go beyond using assumed area relationships to directly map and monitor physical attributes that are critical for habitat quality such as forest structure, which includes canopy cover and height, as well as forest loss and gain dynamics.

Since the early 2000s, regularly updated global forest cover products have been available at a medium to low spatial resolution (≥ 250-m; Hansen *et al.* 2002). However, finer-scale, diffuse forest disturbances that are not detectable with low-resolution sensors (Hansen *et al.* 2008), such as conversion to scattered settlements and farms, can have severe negative consequences for many species, particularly frugivorous species (Johns and Skorupa 1987; Harcourt 1998).

Moreover, deforestation tends to force species to inhabit scattered, remnant forest patches that are embedded in a matrix of human land-uses (Haddad *et al.* 2015). The preservation of remnant habitat patches is a current paradigm in conservation biology (Pimm and Brooks 2013), as well as the preservation or establishment of habitat corridors to link these patches in order to enable the flow of individuals and genes between sub-populations (Bennett 2003). It is therefore paramount to accurately characterise habitat patches and their corridors across a landscape, which is a task that is not suitable for satellite sensors with a coarse spatial resolution (Benson and MacKenzie 1995, Saura 2004, Anteau *et al.* 2014). In order to detect such fine-scale forest loss and landscape heterogeneity, a sensor with a medium to high spatial resolution (< 50-m) is required.

For over four decades, the Landsat series of satellites has been collecting medium-resolution imagery of the Earth's surface. During the majority of this time and for most of the Landsat bands, imagery has been collected at 30-m resolution. While Landsat imagery has existed since 1972, regional to global mapping using Landsat data was not possible. However, as noted in Chapters 1 and 2, three recent advances have transformed the way the scientific community is able to use and process Landsat data to develop a global Landsat-derived forest extent and change data record: (i) systematic global acquisitions, (ii) free and open access to the Landsat archive, and (iii) cloud-computing and data-mining algorithms. The Landsat 7 Enhanced Thematic Mapper Plus (ETM+) sensor, launched in 1999, was the first to implement a systematic global acquisition strategy. Prior to Landsat 7, images were collected inconsistently, limiting the ability to apply a standard algorithm to map global forest cover. Considering free and open access to the Landsat archive, prior to 2008, user acquisition of Landsat imagery required individual purchase, precluding continental- to global-scale analyses. Scientists were limited to the data they could afford rather than the data they needed to conduct land-change analyses. In 2008, the United States Geological Survey (USGS) formally made the Landsat archive open for free access and download to all users (Woodcock *et al.* 2008). Finally, cloud-computing and data-mining algorithms have made it possible to process a large number of Landsat scenes, covering multiple time periods across the globe, as, on an individual basis, a single Landsat image requires significant computing capability to process. The emergence of cloud-based systems for remote sensing applications has, for the first time, helped leverage massive archives of Landsat imagery to map changes in Earth's characteristics over long time spans.

Prior to the opening of the Landsat archive, a prototype data-intensive method for mapping and monitoring forests at a large regional scale using Landsat was developed for the Congo Basin, as part of the United States Agency for International Development (USAID) Central Africa Regional Program for the Environment (CARPE). CARPE is a long-term initiative by USAID to promote sustainable forest management, biodiversity conservation, and climate change mitigation in the Congo Basin by enhancing local, national, and regional natural resource management capacity (http://carpe.umd.edu). Forest extent and change data were a critical input in fulfilling the mission of CARPE; therefore, as part of this program, methods for quantifying forest loss across the entire Congo Basin were prototyped and implemented pre- and post-opening of the archive.

Given the example of CARPE, a global implementation of tree-cover-loss mapping over multiple time periods was deemed feasible. Researchers at the University of Maryland and Google Earth Engine (GEE) teamed up to produce a 'first of its kind' record of forest cover extent and change from 2000 to 2012 (Hansen *et al.* 2013). The global scale, 30-m spatial detail, and annual monitoring interval represented an unprecedented advance in the mapping and monitoring of global forest resources. Now, image processing and classification algorithms can be applied systematically over time as new Landsat data become available, resulting in the production of forest information that is consistent at the global scale and relevant at the local scale.

Annual global updates of the forest-loss layer, including near-real-time alerts, have now been formally incorporated into the World

Resources Institute's Global Forest Watch (GFW) online mapping platform (www.globalforestwatch.org). This tool features an interactive online forest monitoring and alert system that employs the Landsat-derived global forest-cover extent and change data as a primary information layer. GFW's mission is to 'empower people everywhere with the information they need to better manage and conserve forest landscapes'. Users include governments, civil society, and private industry. Applications of the forest cover and loss data vary, and range from countries that utilise GFW in policy frameworks such as the United Nations Framework Convention on Climate Change's (UNFCC) Reducing Emissions from Deforestation and Degradation (REDD+) initiative, to members of the media that report on illegal logging within protected areas. Through GFW, transparency on spatio-temporal trends in forest extent has dramatically increased (Gunther 2015), democratising information on this important global resource. In addition to GFW, the forest cover and change products are freely available for download and analysis at GEE-hosted sites (http://earthenginepartners.appspot.com/ and https://earthengine.google.com). Similar to GFW, GEE is a web-hosted platform that provides users access to Earth observation datasets for free. Additionally, it includes tools to process satellite imagery and conduct custom analyses, from local to global scales, in the cloud, enabling more efficient use of satellite data at broad spatial and temporal scales.

Building capacity for satellite image processing presents a variety of challenges, as it requires advanced technical expertise and substantial financial resources, in the form of personnel, and computer hardware and software (Palumbo *et al.* 2017). This can be a significant burden for conservation organisations on shoestring budgets to bear individually. Free access to objective and accurate information regarding the extent and state of forest habitats can help conservation organisations avoid costs associated with monitoring and evaluation, essential components of successful conservation projects.

The application of global forest cover and change products in the context of forest habitat monitoring has already been demonstrated at regional and global scales. Tracewski *et al.* (2016a) were the first to use the Landsat-based forest cover and loss product developed by Hansen *et al.* (2013) to conduct a global analysis of forest loss in important bird and biodiversity areas (IBAs) using GEE. They found that approximately 2.5 per cent of forest was lost across IBAs between 2000 and 2012, with the highest forest loss rates found in South America and South East Asia. Tracewski *et al.* (2016b) further applied GEE to assess forest loss within the ranges of over 11,000 forest-dependent species. They concluded that extinction risks should be increased for hundreds of species on the International Union for the Conservation of Nature (IUCN) Red List of Threatened Species because of inferred rapid population declines from either habitat loss or because of a restricted area of occupancy owing to the scant forest cover remaining within their range. Results from Tracewski *et al.* (2016b) were used in the most recent update of the IUCN Red List assessment. Their approach will be used in future Red List assessments to rapidly quantify habitat change, which represents a major success of remotely sensed data being used to effect large-scale conservation action. Joshi *et al.* (2016) provided another application of the Hansen *et al.* (2013) global forest cover product, which assessed forest loss within 76 landscapes that were prioritised for tiger conservation, spread across 13 Asian countries. They found that, within these landscapes, 7.7 per cent of forest was lost between 2001 and 2014, much less than was anticipated, but that extensive loss occurred due to the expansion of oil palm plantations. Results of the analysis pointed to the value of satellite-based monitoring of habitat coupled with ground-based observations of tiger occurrence for prioritising the protection of key habitats and corridors. In fact, Indian government officials from states that have tigers are committing to use the GFW platform to monitor performance on achieving the 2020 targets promoted by the Global Tiger Initiative (Joshi *et al.* 2016).

In this chapter, we demonstrate how global forest cover and change products can support conservation monitoring efforts by presenting case studies of habitat change for the endangered chimpanzee

(*Pan troglodytes*) across its range in Africa. First, we provide a brief assessment of habitat loss across the chimpanzee range, using an approach similar to the studies previously discussed. We then focus on the use of remotely sensed data and species modelling to assess change in chimpanzee habitat suitability. Finally, we describe how these analyses and models have been integrated into a dynamic decision-support system to inform ongoing management and conservation decisions at multiple scales.

4.2 MONITORING CHIMPANZEE HABITAT LOSS

Chimpanzees occupy a wide variety of forest and woodland habitats across Central and Western Africa (Figure 4.1). Chimpanzees can serve as an umbrella species (Wrangham *et al.* 2008), meaning protecting their habitat could help conserve habitats for many other species. The chimpanzee has suffered calamitous population declines due to large-scale habitat loss, poaching, disease, and the pet trade (Humle *et al.* 2016). It has been reported that the chimpanzee population may have been reduced by two-thirds over the past 40 years (Kormos *et al.* 2003). An estimated 70 per cent of chimpanzee tropical forest habitats in Africa are now threatened by infrastructure development and land-use change (Nellemann and Newton 2002). Three of the chimpanzee sub-species, Nigeria–Cameroon (*P.t. ellioti*), Central (*P.t. troglodytes*), and Eastern (*P.t. schweinfurthii*) have been classified as endangered on the IUCN Red List, while the Western (*P.t. versus*) sub-species has recently been up-listed to critically endangered (Humle *et al.* 2016).

We assessed the loss of chimpanzee forest habitat using the Hansen *et al.* (2013) tree-cover and loss product and protected areas from the World Database of Protected Areas (WDPA) (www.protectedplanet.net/), a database containing the spatial extent of the world's protected areas (Figure 4.2). We obtained spatial data for country boundaries from the Global Administrative Areas project (http://gadm.org/) to calculate forest loss for each country within the chimpanzee range. The Hansen *et al.* (2013) data are delivered in a geographic coordinate system; therefore, pixel size varies with

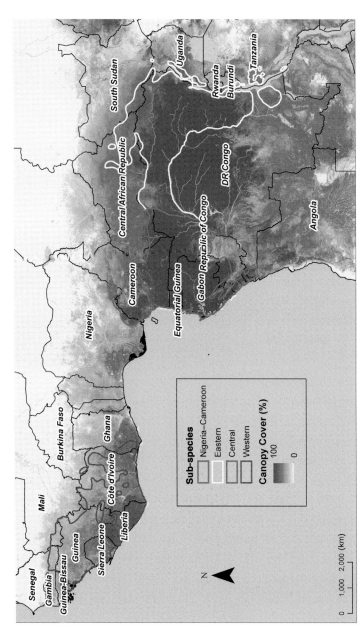

FIGURE 4.I The ranges of the four sub-species of chimpanzee as defir ed by the IUCN Red List are overlaid on a map of per cent tree-canopy cover.

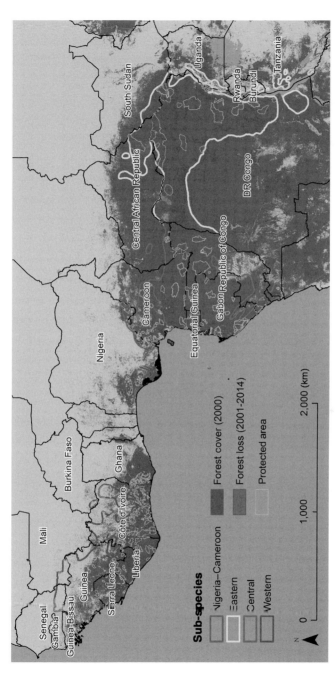

FIGURE 4.2 Forest extent (green) and loss (red) between 2001 and 2014 across the four chimpanzee ranges, with protected areas from WDPA outlined in light blue.
A black and white version of this figure will appear in some formats. For the colour version, please refer to the plate section.)

89

latitude (i.e. the pixel size of the products is approximately 30-m ×
30-m but becomes smaller or bigger when moving toward or away
from the equator, respectively). To estimate the amount of forest and
forest lost, we created a raster layer, representing the area of each pixel
within the study area in square metres.

Numerous definitions of forest exist in the literature. For example,
the Food and Agriculture Organization of the United Nations defines
forest as at least 0.5 hectares of land with trees taller than 5-m and
canopy cover greater than 10 per cent, while the United Nations
Educational, Scientific and Cultural Organization (UNESCO) defines
forest as an area occupied by trees with at least 40 per cent canopy
cover (UNESCO 1973; Food and Agriculture Organization of the
United Nations 2006). For this case study, we defined forest as any
pixel with 25 per cent or more tree-canopy cover as a compromise
between the two definitions. We created a baseline year 2000 forest/non-
forest mask layer by applying a 25 per cent or greater threshold to the
Hansen *et al.* (2013) tree-canopy-cover layer. In other words, pixels with
25 per cent or greater tree canopy cover were considered forested, while
pixels below 25 per cent canopy cover were considered non-forested.
We first calculated the amount of forest in the baseline year 2000 for
each chimpanzee sub-species and disaggregated the results, both for each
country and each protected area contained within each sub-species'
range, in order to quantify the amount of forest loss taking place in
individual countries across the species' ranges and to analyse the effectiveness of their protected areas. For each year up to 2014, we calculated
forest-loss pixels that were located inside our previously defined baseline forest mask. We conducted our analysis on a desktop personal
computer, using the open-source Python 2.7 scripting language.

We found that over 80,000 km^2 of tree cover was lost across
the chimpanzee range between 2001 and 2014. Nearly 9,000 km^2
(11 per cent) of this amount occurred in protected areas. Forest loss was
not distributed equally among the sub-species' ranges. The Eastern and
Western sub-species accounted for the majority of forest loss, with over
34,000 km^2 and 32,000 km^2 of forest loss, respectively (Table 4.1

Table 4.1 Year 2000 baseline forest cover and aggregate forest loss between 2001 and 2014 for the entire range of each chimpanzee sub-species and for the protected areas (PAs) located within each sub-species' range in km².

	Western		Nigeria–Cameroon		Central		Eastern	
	Range	PA	Range	PA	Range	PA	Range	PA
Baseline forest cover (2000)	415,357	74,316	146,211	32,697	689,436	100,064	931,277	142,502
Forest loss (2001-2014)	32,501	6,046	3,671	767	10,692	488	34,046	1,630

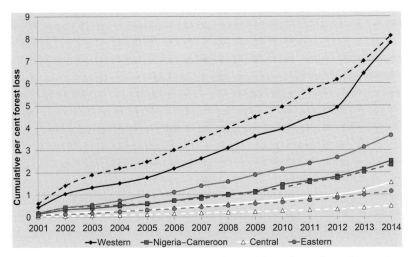

FIGURE 4.3 Cumulative per cent of annual forest loss taking place within each chimpanzee sub-species' range. Solid lines refer to loss within a sub-species' entire range while dashed lines refer to forest loss taking place only within the protected areas located inside each species' range.

and Figure 4.3). The Central and Nigeria–Cameroon sub-species experienced substantially less forest loss, with over 10,000 km^2 and over 3,000 km^2 of forest loss, respectively (Table 4.1). Over two-thirds of the nearly 9,000 km^2 of forest loss that occurred within protected areas was located within the Western chimpanzee range. Moreover, in the Western chimpanzee range, the proportional rate of forest loss within protected areas was higher than that observed across the whole range (Figure 4.3), indicating that forests in protected areas are particularly vulnerable to the removal of trees within the range of this sub-species.

Our results show that forested habitats within the Western chimpanzee range are being lost much faster compared to the other three sub-species (Figure 4.3). The lines that feature steeper slopes between years indicate years with higher rates of change; these are most visible since 2012 especially for the Western sub-species. When comparing results across country boundaries (Figure 4.4), Côte d'Ivoire presents a particularly troubling situation for conservation since a third of its

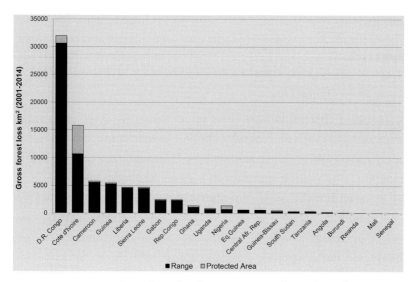

FIGURE 4.4 Gross forest loss between 2001 and 2014 for each country across the chimpanzee range (black bars) and within protected areas (grey bars).

total forest loss has occurred inside protected areas. This could have contributed to a 90 per cent reduction in chimpanzee populations within Côte d'Ivoire as found by Campbell *et al.* (2008). Moreover, some protected areas in the country had a nearly complete removal of forest; Campbell *et al.* (2008) estimated a 93 per cent decline in forest cover in Marahoué National Park in Côte d'Ivoire between 2002 and 2006, which was once thought to support over 900 chimpanzees. These trends paint a bleak picture for the future of the Western sub-species since it was thought that about half of its total population was distributed within Côte d'Ivoire. Guinea and Liberia are now believed to be the remaining population strongholds for this sub-species (Kormos *et al.* 2003, Caldecott and Miles 2005).

There has been increased investment in industrial oil palm production in Africa (Carrere 2013) and recent analysis has shown that a substantial amount of area suitable for oil palm production overlaps with unprotected chimpanzee habitat, especially in Liberia (Wich *et al.* 2014). The expansion of oil palm plantations, or the loss of forest for any other reason, could further impact the long-term persistence of the Western chimpanzee if not implemented in a sustainable manner.

4.3   MODELLING HABITAT SUITABILITY WITH REMOTELY SENSED DATA

Calculating changes in forest loss across the entire range of a species over time is a quick and useful technique to determine the spatial and temporal trends of forest habitat loss. However, this can be insufficient for conservation applications because no single species occupies all areas within the geographic limits of its range (i.e. species are not continually distributed in space; Gaston 1991). Consequently, estimates of change in habitat extent, based solely on information pertaining to the geographic extent of forest habitats across a species' range, will not provide a complete picture of that species' habitat. The explicit incorporation of other environmental parameters is necessary to provide a more accurate representation of a species' preferred habitat. When characterising a species' preferred habitat and prioritising areas for conservation, it is therefore important to consider additional factors such as topographical or climatic conditions, proximity to human activities and threats, or proximity to food sources. Habitat suitability models (HSMs), also referred to as species distribution models (SDMs), or ecological niche models (ENMs), have been applied broadly in the biological sciences literature for assessing the quality of habitat for a species across its range and for identifying the geographic areas that might have the highest likelihood of supporting a species, based on its habitat requirements. In simple terms, habitat consists of the resources and conditions necessary to support the survival and reproduction of a given organism (Hutchinson 1957). It is organism-specific, and characterised as a multi-dimensional function of multiple resources and conditions, each of which can operate at a particular spatial scale (Cushman *et al.* 2010). HSMs often use statistical and machine-learning methods to determine empirical relationships between observed species' occurrences and environmental descriptors, such as resource, biotic, and climatic factors, in order to provide a quantitative representation of a suitable habitat. HSMs can serve a variety of purposes, including the identification of areas of high conservation potential, the identification of sites for species reintroductions, the design of wildlife corridors, the prediction of sites at risk for disease or pest species outbreaks or exotic species invasions, and the prediction of potential changes in species distributions in response to land-use and climate change (Guisan *et al.* 2000, Manel *et al.* 2001, Stickler and Southworth 2008).

Results from HSMs are most convincing when they are calibrated with both presence and absence data. However, due to the difficulty of acquiring ecologically meaningful absence data, HSMs are often constructed with observation-based 'presence-only' data (Hastie and Fithian 2013). Verifiable true absences are difficult to collect, especially when a species of interest is relatively rare and tends to wander about. For example, a biological survey could be conducted in a particular location that may be considered suitable for a particular species, but no individual of the species was present at the time the surveys were conducted, or the species was present but went undetected. To address this issue, researchers can utilise so-called 'pseudo-absences' or background points that are chosen through a heuristic approach or at random, but are treated in the same way as true absence points (Barbet-Massin *et al.* 2012). When models are fitted with presence-only data, the output can be interpreted as a relative measure of environmental suitability rather than the probability of presence (Fitzpatrick *et al.* 2013, Merow *et al.* 2013).

To date, only a handful of attempts have been made to model chimpanzee habitat suitability in order to answer questions relevant to chimpanzee conservation and ecology. These were either modelled at a low spatial resolution (≥ 1-km), covered a small portion of the chimpanzee range, or depicted habitat suitability for just a single historical point in time. To be most relevant for decision support, information on the habitat condition should be based on current data and easily updated as new information becomes available. The most recent advance in this direction was developed by Jantz *et al.* (2016) in collaboration with the Jane Goodall Institute (JGI). A low-resolution (5-km) chimpanzee habitat suitability map produced by Junker *et al.* (2012), the first study to map range-wide chimpanzee habitat suitability, served as the foundation for the approach developed by Jantz *et al.* (2016). Junker *et al.* (2012) used climate, human impact, and vegetation variables as model inputs to map habitat suitability for all four chimpanzee sub-species for the 1990s and 2000s time periods. While Junker *et al.* (2012) created habitat suitability maps for two time periods, only the gridded estimates of human population density changed between the 1990s and 2000s time periods. Consequently, any modelled change in the extent of suitable habitat was entirely due to changes associated with human population density.

Results from the approach detailed in Jantz *et al.* (2016) consisted of mapped data of chimpanzee habitat suitability, produced at a finer spatial resolution, around a dynamic decision-support system that enables systematic updates as new data become available. This will inform ongoing conservation. To this end, Jantz *et al.* (2016) developed a suite of remotely sensed variables, including metrics of forest structure (i.e. forest canopy cover and height), forest disturbance, and topography in order to identify indicators of chimpanzee habitat suitability, compatible with those used in Junker *et al.* (2012). Remotely sensed variables developed by Jantz *et al.* (2016) and covering the 2000–2003 time period were coarsened to the same 5-km spatial resolution as the Junker *et al.* (2012) habitat suitability map, and data were extracted for all grid cells. Data were selected for this specific time period in order to have a close temporal match with the Junker *et al.* (2012) habitat suitability map. Random Forests regression was then used to determine which combination of remotely sensed variables could best predict habitat suitability. Random Forests is an ensemble technique in the Classification and Regression Trees (CART) family of machine-learning methods and has the attractive properties of being able to capture non-linear relationships and being insensitive to non-normally distributed variables and correlated predictor variables (Breiman *et al.* 1984, Breiman 2001). The remotely sensed metrics served as the independent variables and the Junker *et al.* (2012) habitat suitability map served as the dependent variable.

The Random Forests regression model explained 82 (±0.2) per cent of the variance in the Junker *et al.* (2012) suitability map. Elevation, Landsat ETM+ band 5 (a measure of forest structure), and Landsat-derived per cent canopy cover were found to be the best predictors of chimpanzee habitat suitability. An important assumption associated with the modelling approach was that habitat suitability was insensitive to changes in spatial resolution; that is, the model assumed that the relationship between the remotely sensed metrics and habitat suitability is the same at both the 5-km and 30-m resolutions. Previous studies that have demonstrated that tree-canopy cover and spectral reflectance (both of which were found to be strong predictors of chimpanzee habitat suitability) are insensitive to changes in resolution (Gao *et al.* 2006, Sexton *et al.* 2013). Moreover, all variables used as input into the Random Forests model were of a continuous nature and the favourable scaling properties of continuous data have been known for a long time (Stevens 1946).

Range-wide maps of habitat suitability can be created at 30-m resolution with this approach (Figure 4.5A). Its main advantage is its reliance on remotely sensed forest data, which can be readily updated to enable ongoing monitoring. The majority of the data used as input into the Random Forests model is from the Landsat 7 ETM+ sensor, which, along with the recently launched Landsat 8 satellite, continues to acquire imagery. The University of Maryland is committed to processing this imagery and releasing annual updates of forest loss and gain dynamics into the indefinite future, as long as systematic acquisition continues. As new data are made available, the remotely sensed metrics can be updated and new predictions of habitat suitability can be made in order to reveal changes in habitat over time. For example, Figure 4.5B shows the impact of a selective logging operation on chimpanzee habitat suitability in the Republic of the Congo between 2003 and 2012, while Figure 4.5C shows the loss of

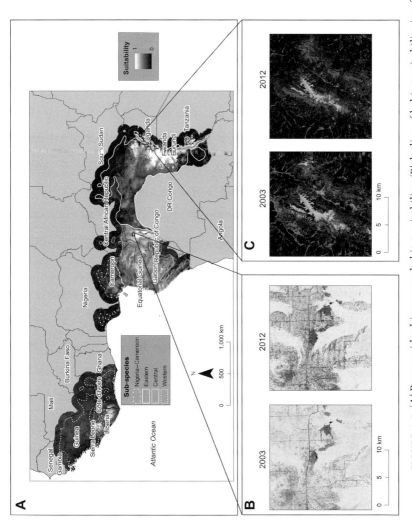

FIGURE 4.5 (A) Range-wide chimpanzee habitat suitability; (B) decline of habitat suitability in a forest in the Republic of Congo, experiencing selective logging; and (C) loss of habitat patches in the Budongo–Bugoma corridor, Uganda.

habitat in a corridor connecting Bugoma and Budongo National Parks in Uganda over the same period, primarily due to agricultural encroachment (McLennan 2008).

4.4 CONVERTING PIXELS INTO CONSERVATION DECISIONS

The JGI is currently incorporating the approaches described above into their monitoring and decision-making activities. By protecting chimpanzees and inspiring action to conserve the natural world, the JGI's mission is to improve the lives of people, animals, and the environment. The institute achieves this objective by building the capacity of local communities, protected-area rangers, and governments to better manage their natural resources, including by leveraging innovative technology solutions to empower community-led conservation efforts across the chimpanzee range. The JGI's main conservation objective is to stabilise and protect wild chimpanzee populations by addressing the most important threats to their survival, such as habitat loss.

Since 2000, the JGI has used satellite data derived from Landsat sensors as well as from other medium- to high-resolution sensors to monitor the change in chimpanzee habitats and human land-use and inform conservation planning (Pintea *et al.* 2003, Pintea 2005, 2007, Pusey *et al.* 2007). Starting in 1994, tree planting was one of the main JGI community conservation strategies to address forest and chimpanzee habitat losses on village lands outside Gombe National Park in Tanzania. This was based on the assumption that local communities would not cut down trees if alternative sources of firewood and timber were available. However, Landsat observations from 1972 and 1999, combined with 2001 1-m resolution commercial Ikonos satellite images, revealed that firewood collection and logging were not the main risks to chimpanzee habitat loss. Rather, conversion of forests and woodlands to farms and settlements were the most important threats to chimpanzee habitats outside Gombe. Therefore, it was concluded that while tree planting and nurseries could still be an important strategy to support people's livelihoods and need for firewood and timber, additional strategies were required in order to stop chimpanzee habitat loss occuring outside the national park. The JGI thus created a new strategy to develop and implement participatory village land-use plans and establish zones for agriculture and community-managed village forest reserves that would benefit watersheds, people, and chimpanzees. This new strategy has worked in many villages, with natural forest regeneration occurring, confirmed by DigitalGlobe satellite imagery between 2005 and 2014 (Pintea 2016). Satellite imagery are used continually by the JGI to assess the enforcement of village land-use plans and village forest reserves by monitoring forest loss and reforestation. However, due to limited capacity and resources, to date, the JGI has only been able to analyse selected Landsat scenes on a site-by-site and per-project basis, with limited temporal and geographic coverage. The lack of standardisation of the satellite-derived data and information products across geographic space and time precluded a comparison of chimpanzee habitat conditions across the entire species' range. The JGI thus recognised this as a critical issue that, if resolved, could provide their staff and partners with a more consistent understanding of the status of chimpanzee populations across the species' entire range, in turn enabling the JGI to more cost-effectively develop and evaluate their conservation strategies at different spatial scales.

With this in mind, the JGI partnered with the University of Maryland (UMD), the National Aeronautics and Space Administration (NASA), and others to develop a decision-support system (DSS) to monitor and forecast chimpanzee habitat health and connectivity. The system uses 15 years of Earth observations from 30-m resolution Landsat data as input into the habitat suitability model described in Section 4.3 above (see Jantz et al. (2016) for more detail) and a model forecasting potential future land-use change. This effort was further enhanced by crowd-sourced field data, collected from a variety of sources, including those from local communities and rangers, using mobile smartphones and tablets.

The DSS is an analytical tool running in the JGI's Microsoft Azure cloud, which JGI staff can use to easily update chimpanzee habitat metrics as new Landsat imagery, better suitability models, and additional chimpanzee presence data become available, to provide timely information about chimpanzee habitats across a range of spatial scales. Resulting data layers derived from these habitat metrics are then used by JGI managers to inform the JGI's institutional strategies. These layers are also shared with partners to support specific conservation planning efforts at the local, national, or regional scales across chimpanzee ranges in Africa.

An interface to the DSS is being developed using ESRI's Arc GIS platform, which consists of dynamic web dashboards that will allow decision-makers to retrieve and communicate model results through data visualisations and online web maps. This information is intended to be used in support of developing conservation strategies using Open Standards for the Practice of Conservation (OSPC). OSPC is a collaborative and adaptive management process (Conservation Measures Partnership 2013) that the JGI has been using since 2005 in order to help improve the JGI's conservation strategies and measure success in a manner that enables adaptation and learning over time. The OSPC management framework helps define criteria and thresholds for converting gridded remote sensing data into information potentially useful to inform decision-making. Indeed, it was the OSPC process that helped the JGI formulate their adaptive management strategies to assist more effective conservation planning for chimpanzee protection around Gombe National Park, as described above. This OSPC approach is an ongoing process, which engages critical stakeholders, including local and national governments, communities, academia, international organisations such as United Nations–Great Apes Survival Partnership (GRASP), IUCN Primate Specialists Group, and other non-governmental organisations, through a series of diagnostic steps that culminate in the development and implementation of clearly defined objectives and strategic actions. Systematic use of OSPC allows the JGI to learn what works, what does not, and why, and to adapt and improve the effectiveness of conservation efforts.

The DSS was designed to support the OSPC adaptive management process through the provision of continuously updated consistent remote sensing data and models. The JGI staff upload the latest Landsat datasets from the University of Maryland into the JGI's Microsoft Azure cloud account and run automatic routines to systematically calculate updated habitat metrics, such as habitat suitability and change. Work is in progress to develop interactive dashboards using ESRI's Arc GIS platform that managers and stakeholders can use during collaborative conservation planning workshops, to discuss and agree upon how to interpret DSS habitat metrics. Important steps in the OSPC decision-making process, where satellite data are critical, are a viability and threats assessment, as well as determining how conservation success should be defined and measured. This entails quantifying changes in chimpanzee populations and habitats over time, which can be achieved through four steps: (1) determining the current status of chimpanzee populations and habitat condition, (2) defining what a desired healthy state might look like, (3) identifying critical threats to that desired state, and (4) determining a mechanism to measure the efficacy of targeted interventions.

The DSS uses annually updated Landsat satellite datasets to derive a number of temporally dynamic indices that inform various management questions, as defined by specific planning processes using OSPC standards. For example, in Tanzania, the DSS has supported a second iteration of the Gombe–Mahale Ecosystem Plan and the development of a national chimpanzee management plan (Tanzania Wildlife Research Institute, unpublished draft report, 2017). Remotely sensed datasets are used within the DSS to measure two indicators of chimpanzee habitat viability. The first indicator relates to the area of forest. The key ecological attribute (KEA) for this indicator is the per cent of forest cover loss between 2001 and 2014, within the areas of forest and woodland that were considered suitable for chimpanzees in 2000 (Figure 4.6A). The second indicator relates to habitat condition. Here, the KEA was forest cover loss occurring between 2001 and 2014, within evergreen and riverine forests identified as potentially suitable for chimpanzees (Figure 4.6B).

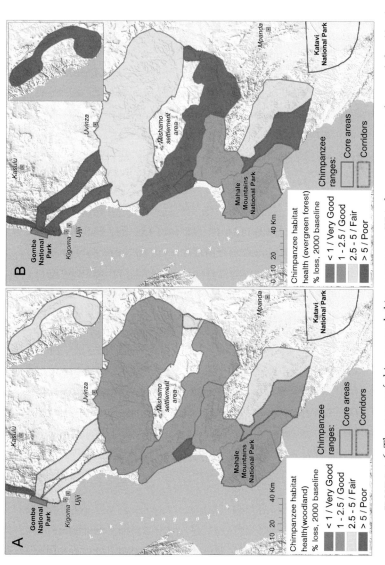

FIGURE 4.6 The status of chimpanzee habitat across two KEAs by management polygons, defined by the chimpanzee core areas and corridors. (A) A size-category KEA, measured as per cent forest and woodland loss within a suitable habitat. (B) A condition KEA, measured as per cent evergreen forest loss within a suitable habitat. Conservation action planning stakeholders agreed to interpret chimpanzee habitat health as Very Good (dark green) if per cent forest loss was less than 1 per cent, Good (light green) as 1–2.5 per cent, Fair as 2.5–5.0 per cent and Poor if more than 5 per cent was lost compared to the 2000 baseline. (A black and white version of this figure will appear in some formats. For the colour version, please refer to the plate section.)

Evergreen and riverine forests are naturally scarce and represent only around 2 per cent of the forest cover in the chimpanzee range in Tanzania. However, these narrow and small patches of forest are critical for the viability of chimpanzees in the region. Chimpanzees living in dry forest–savanna–woodland environments, as in Tanzania, are adapted to use dry miombo woodlands, but their range always includes evergreen forests. Additional KEAs, such as chimpanzee nest and population density, have been identified as important indicators for assessing chimpanzee habitat. While chimpanzee nests could be detected using unmanned aerial vehicles (UAVs), no direct signs of chimpanzee presence have yet been directly monitored from any satellite remote sensing platform, and, therefore, these last two KEAs are not currently included as part of the DSS.

Managers are not making decisions at the scale of individual pixels. Consequently, they need information at the level of management areas. Accordingly, the DSS was designed to use the OBPC process to convert pixels of data into useful management information for specific management units. In the case of Tanzania, habitat condition indicators were estimated in different management polygons to inform stakeholder dialogue and decisions, regarding village and governmentally managed protected areas, land tenure and land-use, and chimpanzee core habitats and corridors.

Remote sensing data have revealed that conservation efforts in the Gombe–Mahale Ecosystem have had mixed success. Conservation strategies such as village land-use planning have been successful in protecting miombo woodland chimpanzee habitats. However, these strategies have been less useful in protecting riverine and evergreen forests as measured by the KEA indicator on habitat condition. Riverine forests are patchy and narrow, and are often smaller than 15-m wide. To visually validate and communicate to decision-makers the results of the DSS assessment, we used very high resolution commercial satellite images at 60-cm (provided in-kind and free of charge by DigitalGlobe to the JGI) to zoom into the riverine forest-loss hotspots that are detected by the DSS. These higher-resolution satellite images indicated that agriculture had expanded into more remote and difficult-to-protect pristine forest regions, often located in riverine areas, resulting in the loss of evergreen forests in particular. Informed by the DSS and OSPC process, decision-makers agreed to prioritise the adoption of new or modified conservation strategies, to increase the protection of these remote evergreen and riverine forests, and, in turn, to improve the long-term survival of chimpanzees in Tanzania.

Initially, we assumed that one index of chimpanzee habitat health would be capable of meeting most of the decision-makers' needs. However, we learned that this was too simplistic. The OSPC process requires multiple indicators of chimpanzee habitat status and threats, which could not be consolidated into one index. For example, during the viability assessment stage, three categories of KEAs that have to be defined: 'size', which is a measure of the area of chimpanzee habitat; 'condition', a measure of the composition and structure that characterise the habitat (e.g. habitat dominated by evergreen forest or miombo woodlands and savanna); and 'landscape context', which is an assessment of chimpanzee habitat environment, such as distance from human settlements or connectivity. Different combinations of remote sensing products, such as per cent canopy cover, chimpanzee suitability, and forest loss data were used to define specific indicators within the three categories. Breaking down the index into smaller indicators, and allowing decision-makers to assess various components of habitat viability and threats individually, was important to build a deeper understanding of the drivers of chimpanzee habitat health and the associated sources of threats, ultimately improving conservation decisions. In the case of the Greater Gombe Ecosystem, Landsat Multispectral Scanner (MSS) and ETM+ data showed the extent of forest loss between 1972 and 1999. Monitoring Forest loss increased awareness that chimpanzee habitat outside the park needed conservation intervention. However, forest-loss data alone was not actionable and could not be used to inform what conservation strategy or actions could potentially minimise or stop deforestation.

The JGI was able to identify better conservation strategies when they combined forest loss data, OSPC conceptual models, Ikonos imagery, and field surveys. From this, they realised that the main habitat threat was the conversion of forests to subsistence agriculture. Addressing threats to forests from subsistence agriculture requires different strategies than those from logging or firewood collection. Therefore, we developed the idea to integrate KEAs in the DSS within the OSPC framework, to allow for multiple ways of dissecting and analysing the remotely sensed habitat data and to convert it to actionable information.

We also assumed that the DSS would be most helpful if it showed the maps and data as continuous variables at 30-m resolution. However, decision-makers struggled to convert 30-m maps into information relevant to management. High-spatial-resolution maps are important to accurately assess heterogeneous chimpanzee habitats and threats, such as small, scattered, and patchy farms in narrow riverine forests. However, managers do not make decisions at the pixel or individual-farm scales. Instead, they make decisions at the level of the polygons or management units for which they are responsible. These management units could be hierarchical and nested, depending on the scale of the conservation planning process. For example, a village forest reserve could be too small for decision-makers to consider when planning at the national scale, but it could be nested into the larger ecosystem boundary that informs planning decisions. We therefore decided to add a tool to the DSS that enables decision-makers to overlay maps of continuous remote sensing indicators on selected management areas, and create categorical variables from the indicator maps by interactively setting thresholds to define states of 'Very Good', 'Good', 'Fair', or 'Poor', in accordance with OSPC methodology.

A key lesson learned during the chimpanzee habitat health DSS project was that we need quality remote sensing products that address specific information needs (Pintea *et al.* 2002) as well as management frameworks for decision-makers, so they can discuss and agree how to interpret these data. In places where there are no management plans or policy guidelines, OSPC could be used as a management framework that guides decision-makers through the process of setting indicators and standards. In the case of Tanzania, the availability of a country-wide 30-m forest-cover dataset that detected loss in dry forests and miombo woodlands was essential. However, what made the data actionable was that stakeholders using the OSPC process discussed and agreed on thresholds of how much forest loss within a management polygon indicates 'Very Good' and 'Good' habitat health versus what should be interpreted as 'Fair' and 'Poor' condition, thus identifying which existing conservation strategies should be improved. These KEAs were created and updated over the course of several workshops, as part of the adaptive OSPC management cycle.

The vision is to scale our process up to the entire chimpanzee range. To this end, the JGI is now working with the Conservation Measures Partnership, Foundations of Success (FOS), IUCN, GRASP, NASA, the US Fish and Wildlife Service, and other partners, to develop a draft standard framework for converting spatial conservation data into meaningful management information, using chimpanzee conservation in Africa as an initial test case. The partners first convened at the IUCN World Conservation Congress in 2016 in Honolulu. Stakeholders agreed to set an interdisciplinary working group of conservation practitioners, researchers, funders, and policy-makers to pilot-test integrating spatial data with the OSPC in the context of large-scale conservation projects around the world. The results of these efforts were presented as part of a symposium organised by the JGI and FOS at the Society for Conservation Biology's 28th International Congress in Cartagena, in 2017.

One final lesson that has been learnt is the importance of working closely with protected-area and local district and natural resource managers. This enabled us to jointly assess their information needs and identify barriers to information use, such as a lack of resources or technical capacity to use remote sensing data. Most of the local communities and government officers have limited geographic information system (GIS) capacity and resources to maintain and update such geospatial databases. Therefore, in addition to developing remote sensing products, the JGI has been facilitating access and building local capacity to use geospatial technologies. These capacity-building efforts have played an important role in enabling remote sensing data to inform participatory village land-use plans outside Gombe National Park, and have had an input into shaping future landscapes that better work for people and chimpanzees.

4.5   CONCLUSIONS AND OUTLOOK

We have presented two applications that demonstrate the use of freely available satellite data and primarily open-source software for supporting conservation objectives. Both use data for monitoring species' populations to assist the evaluation of conservation actions and provide qualitatively different information and perspectives on habitat condition. The first approach quantifies forest loss for a particular area of interest (in this case within a species' range). This approach is relatively simple to implement, and the results can be easily communicated by reporting the area of forest loss occurring over a particular time period, or by reporting the proportion of forest loss compared to a previously defined baseline of forest extent. Although our particular analysis was undertaken using a combination of Python code and GIS software, these tools are not a prerequisite for other users who wish to conduct similar analyses but who might lack these skills or tools. Custom analyses can be conducted within the GFW web-based tool, by delineating an area of interest using the on-screen mapping tool or by uploading geographic coordinates for a predefined area. All computations are processed in the cloud, over a user-defined time period. Results can be instantly viewed and downloaded, and shared and integrated into professional reports.

The second application measures the quality of chimpanzee habitat, demonstrated here via models developed by Jantz *et al.* (2016). This application is somewhat more computer-intensive and analytically demanding, but provides an arguably more accurate depiction of habitat, by determining empirical relationships between species observations and measures of environmental conditions. This model-based result of relative habitat suitability is spatially explicit and therefore can be visualised in map form. Maps like these can be valuable for spatially prioritising regions with the highest suitability for conservation action, in addition to identifying potentially suitable, but unsurveyed, areas that may be important for a future survey effort (Hickey *et al.* 2013). In addition, habitat suitability models provide a better understanding of which environmental conditions might be the largest drivers that are determining the observed distribution of a species. One aspect, which is potentially of great importance for conservation managers, is that model outputs can be extremely valuable to developing long-term monitoring plans, especially if the models' environmental variables are based on regularly updated, remotely sensed data, as in Jantz *et al.* (2016). This enables timely updates of status and systematic near-real-time monitoring of changes to habitats over time. The ability to systematically detect and dynamically visualise these changes can provide conservation managers with critical information for prioritising vulnerable habitats and measuring the success of conservation actions.

Looking to the future, an exciting extension of habitat suitability modelling is the more recent efforts to improve the accuracy of population abundance modeling (Ashcroft *et al.* 2017). Since most conservation managers require information on a species' population size in order to develop strategies for maintaining healthy and viable wild populations in the long term, this is a significant advance. Another emerging area of investigation strives to link observed species' population numbers with pertinent environmental information to estimate population densities across relatively large regions. Plumptre *et al.* (2015) provided a proof of concept for gorillas and chimpanzees in eastern Democratic Republic of the Congo.

Importantly, remote sensing data such as those used here can aid in timely intervention in situations where forests are unlawfully removed, such as unpermitted timber extraction or mining. In a perfect world, sufficient rangers or other enforcement officers would be able to patrol a protected area and intervene in these situations. However, institutional enforcement of laws that protect wildlife and forests is often weak or non-existent in tropical developing countries, which lack the resources to fund these efforts adequately (Bruner *et al.* 2001). As part of the GFW initiative, remotely sensed data are quickly processed and enable the timely detection of forest loss. Current auxiliary products that complement the Hansen *et al.* (2013) data and are featured on the GFW website include: Terra-i alerts, which are updated on a monthly time-scale and are mapped at 250-m resolution across the tropics (Reymondin *et al.* 2012); Sistema de Alerta de Desmatamento (SAD) alerts, which are also updated monthly and mapped at 250-m but are available for the Brazilian Amazon only (De Souza *et al.* 2009); and, most recently, the Global Land Analysis and Discovery (GLAD) laboratory at the University of Maryland released Landsat-based alerts, updated weekly at 30-m resolution, across the tropics (Hansen *et al.* 2016). Using the GFW web-based tool, users can delineate an area of interest on the GFW map and subscribe to these alerts, and they receive a notification whenever deforestation is detected in their area of interest. They can also use Forest Watcher, a free mobile app to download and then, offline, locate, validate, and report forest loss alerts in the field (Petersen and Pintea 2017).

Traditionally, ecological sites are monitored using ground-based field surveys, which are met with some of the challenges mentioned earlier, in both cost and efficiency. On top of this, field surveys conducted in many long-term ecological sites tend to be selected for their pristine nature and therefore can be spatially biased. For example, Dornelas *et al.* (2014) did not find evidence of systematic species loss from an assemblage of long-term ecological field studies. This finding contradicted the well-accepted perspective that there is an ongoing conservation crisis and that the planet is losing species at a rate

1,000 times faster than expected, because of human activities (Pimm *et al.* 2014). This discrepancy suggests the sites selected for long-term ecological study are located in areas that are not experiencing substantial human disturbance. Remote sensing data allow for more extensive areas to be monitored and can provide information from multiple remotely sensed products, such as the three Landsat-based data products featured in Figure 4.7B, for an area near Kisangani, Democratic Republic of the Congo. Areas in bright green represent primary forest that is relatively distant from human activities, yellow to red hues represent the year that forest loss occurred, and darker to lighter cyan tones represent lower to higher human population densities, respectively. In order to accurately determine population trends for numerous species, the allocation of new, long-term ecological field sites could incorporate synoptic geospatial data, such as those shown in Figure 4.7, with the goal of developing efficient, design-based scientific studies enabling inferences from probability-based field data at a large scale.

Satellite data products complement existing ground-based data and improve our understanding of chimpanzee habitat condition. The chimpanzee range spans a massive area and includes regions of dense tropical forest, making ground-based habitat monitoring intractable. Indicators of habitat condition that are generated from systematically acquired satellite data will enable conservation managers to more efficiently monitor habitat change across the entire chimpanzee range. Here, we have demonstrated the immense utility of satellite-derived data to build a dynamic tool that enables consistent, comparable analysis of chimpanzee habitat change over long time spans. Since these data are systematically generated and hosted on a cloud-based platform, conservation managers do not need to spend time learning how to run the model; instead, they benefit from data and maps that are summarised by the management units and are directly relevant to their decision-making. The development of this type of dynamic decision-making tool can be extremely valuable for identifying sites to target conservation

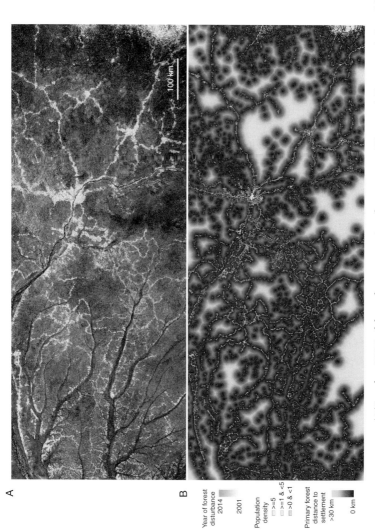

A

B

Year of forest
disturbance
2014

2001

Population
density
>=5
>=1 & <5
>0 & <1

Primary forest
distance to
settlement
>30 km

0 km

FIGURE 4.7 (A) Landsat ETM+ false colour composite of an area near Kisangani, Democratic Republic of the Congo. (B) A map of the same geographic area shows the relative distance from primary forest to human settlements (brighter green areas are located farthest away from human settlements), forest disturbance from 2001 to 2014 shown in a gradient from yellow to red, and population density shown in cyan tones. (A black and white version of this figure will appear in some formats. For the colour version, please refer to the plate section.)

112

prioritisation activities and to better evaluate conservation actions over time.

ACKNOWLEDGEMENTS

Financial support for the work presented in this chapter comes from a grant awarded under NASA's Ecological Forecasting for Conservation and Natural Resource Management program (NNX14AB96 G) and the USAID CARPE program (NNX14AR46 G) and with support of the American people through the USAID under the terms of Co-operative Agreement No. AID-621-A-00-10-00009 – WCS, and through a sub-award with International Resources Group under the Promoting Tanzania's Environment, Conservation and Tourism (PROTECT) Project. Special thanks to DigitalGlobe, Esri, and Microsoft for in-kind contribution and support with satellite imagery, GIS software, and cloud storage and computing costs.

REFERENCES

Anteau, M. J., Wiltermuth, M. T., Sherfy, M. H., and Shaffer, T. L. (2014). Measuring and predicting abundance and dynamics of habitat for piping plovers on a large reservoir. *Ecological Modelling*, **272**, 16–27.

Ashcroft, M. B., King, D. H., Raymond, B., *et al.* (2017). Moving beyond presence and absence when examining changes in species distributions. *Global Change Biology*, **23**, 2929–2940.

Avissar, R. and Werth, D. (2005). Global hydroclimatological teleconnections resulting from tropical deforestation. *Journal of Hydrometeorology*, **6**, 134–145.

Barbet-Massin, M., Jiguet, F., Albert, C. H., and Thuiller, W. (2012). Selecting pseudo-absences for species distribution models: how, where and how many? *Methods in Ecology and Evolution*, **3**, 327–338.

Bennett, A. F. (2003). *Linkages in the Landscape: The Role of Corridors and Connectivity in Wildlife Conservation*. Gland: IUCN.

Benson, B. J. and MacKenzie, M. D. (1995). Effects of sensor spatial resolution on landscape structure parameters. *Landscape Ecology*, **10**, 113–120.

Breiman, L. (2001). Random Forests. *Machine Learning*, **45**, 5–32.

Breiman, L., Friedman, J., Stone, C. J., and Olsen R. A. (1984). *Regression Trees*. Washington DC: Chapman and Hall/CRC.

Brook, B., Bradshaw, C. J. A., Koh, L. P., and Sodhi N. S. (2006). Momentum drives the crash: mass extinction in the tropics. *Biotropica*, **38**, 302–305.

Brooks, T. M., Pimm, S. L., and Oyugi, J. O. (1999). Time lag between deforestation and bird extinction in tropical forest fragments. *Conservation Biology*, **13**, 1140–1150.

Bruner, A. G., Gullison, R. E., Rice, R. E., and da Fonseca, G. A. B. (2001). Effectiveness of parks in protecting tropical biodiversity. *Science*, **291**, 125–128.

Caldecott, J. and Miles, L. (2005). *World Atlas of the Great Apes and Their Conservation*. Cambridge: UNEP World Conservation Monitoring Centre.

Campbell, G., Kuehl, H., Kouamé, P. N. G., and Boesch C. (2008). Alarming decline of West African chimpanzees in Côte d'Ivoire. *Current Biology*, **18**, R903–R904.

Carrere, R. (2013). Oil palm in Africa: past, present, and future scenarios. World Rainforest Movement. See http://wrm.org.uy/wp-content/uploads/2014/08/Oil_Palm_in_Africa_2013.pdf.

Conservation Measures Partnership (2013). Open standards for the practice of conservation; version 3.0. See http://cmp-openstandards.org/wp-content/uploads/2014/03/CMP-OS-V3-0-Final.pdf.

Cushman, S. A., Gutzweiler, K., Evans, J. S., and McGarigal, K. (2010). The gradient paradigm: a conceptual and analytical framework for landscape ecology. In Cushman, S. A. and Huettmann, F., eds., *Spatial Complexity, Informatics, and Wildlife Conservation*. Tokyo: Springer Verlag, pp. 83–108.

Defries, R. (2012). Why forest monitoring matters for people and the planet. In Achard, F. and Hansen, M. C., eds., *Global Forest Monitoring from Earth Observation*. Boca Raton, FL: CRC Press, pp. 1–12.

De Souza, C. M., Hayashi, S., and Veríssimo, A. (2009). Near real-time deforestation detection for enforcement of forest reserves in Mato Grosso. See http://77.243.131.160/pub/fig_wb_2009/papers/trn/trn_2_souza.pdf.

Dornelas, M., Gotelli, N. J., McGill, B., *et al.* (2014). Assemblage time series reveal biodiversity change but not systematic loss. *Science*, **344**, 296–299.

Fitzpatrick, M. C., Gotelli, N. J., and Ellison, A. M. (2013). MaxEnt versus MaxLike: empirical comparisons with ant species distributions. *Ecosphere*, **4**, article 55.

Food and Agriculture Organization of the United Nations (2006). *Global Forest Resources Assessment 2005, Main Report. Progress towards Sustainable Forest Management*. Rome: Food and Agriculture Organization of the United Nations.

Gao, F., Masek, J., Schwaller, M., and Hall, F. (2006). On the blending of the MODIS and Landsat ETM+ surface reflectance. *IEEE Transactions on Geoscience and Remote Sensing*, **44**, 2207–2218.

Gaston, K. J. (1991). How large is a species' geographic range? *Oikos*, **61**, 434–438.

Guisan, A. and Zimmermann, N. E. (2000). Predictive habitat distribution models in ecology. *Ecological Modelling*, **135**, 147–186.

Gunther, M. (2015). Google-powered map helps fight deforestation. *The Guardian*, 10 March. See www.theguardian.com/sustainable-business/2015/mar/10/goo gle-earth-engine-maps-forest-watch-deforestation-environment.

Haddad, N. M., Brudvig, L. A., Clobert, J., *et al.* (2015). Habitat fragmentation and its lasting impact on Earth's ecosystems. *Science Advances*, **1**, 1–9.

Hansen, M. C., DeFries, R. S., Townshend, J. R. G., *et al.* (2002). Towards an operational MODIS continuous field of percent tree cover algorithm: examples using AVHRR and MODIS data. *Remote Sensing of Environment*, **83**, 303–319.

Hansen, M. C., Roy, D. P., Lindquist, E., *et al.* (2008). A method for integrating MODIS and Landsat data for systematic monitoring of forest cover and change in the Congo Basin. *Remote Sensing of Environment*, **112**, 2495–2513.

Hansen, M. C., Potapov, P. V, Moore, R., *et al.* (2013). High-resolution global maps of 21st-century forest cover change. *Science*, **342**, 850–853.

Hansen, M. C., Krylov, A., Tyukavina, A., *et al.* (2016). Humid tropical forest disturbance alerts using Landsat data. *Environmental Research Letters*, **11**, 34008.

Harcourt, A. (1998). Ecological indicators of risk for primates, as judged by species' susceptibility to logging. In Caro, T., ed. *Behavioral Ecology and Conservation Biology*. Oxford: Oxford University Press, pp. 56–79.

Hastie, T. and Fithian, W. (2013). Inference from presence-only data: the ongoing controversy. *Ecography*, **36**, 864–867.

Hickey, J. R., Nackoney, J., Nibbelink, N. P., *et al.* (2013). Human proximity and habitat fragmentation are key drivers of the rangewide bonobo distribution. *Biodiversity and Conservation*, **22**, 3085–3104.

Hilton-Taylor, C., Pollock, C. M., Chanson, J. S., *et al.* (2008). State of the world's species. In Vie, J.-C., Hilton-Taylor, C., and Stuart, S. N., eds. *Wildlife in a Changing World: An analysis of the 2008 IUCN Red List of Threatened Species*. Gland: IUCN, pp. 15–41.

Humle, T., Maisels, F., Oates, J. F., Plumptre, A. J., and Williamson, E. A. (2016). Pan troglodytes. *The IUCN Red List of Threatened Species 2016*. See www .iucnredlist.org/details/15933/0.

Hutchinson, G. (1957). Concluding remarks. *Cold Spring Harbor Symposia on Quantitative Biology*, **22**, 415–427.

Jantz, P., Goetz, S., and Laporte, N. (2014). Carbon stock corridors to mitigate climate change and promote biodiversity in the tropics. *Nature Climate Change*, **4**, 138–142.

Jantz, S. M., Pintea, L., Nackoney, J., and Hansen, M. C. (2016). Landsat ETM+ and SRTM data provide near real-time monitoring of Chimpanzee (*Pan troglodytes*) habitats in Africa. *Remote Sensing*, **8**, 1–16.

Johns, A. D. and Skorupa, J. P. (1987). Responses of rain-forest primates to habitat disturbance: a review. *International Journal of Primatology*, **8**, 157–191.

Joshi, A. R., Dinerstein, E., Wikramanayake, E., *et al.* (2016). Tracking changes and preventing loss in critical tiger habitat. *Science Advances*, **2**, 1–8.

Junker, J., Blake, S., Boesch, C., *et al.* (2012). Recent decline in suitable environmental conditions for African great ape. *Diversity and Distributions*, **18**, 1077–1091.

Kim, D., Sexton, J. O., and Townshend, J. R. (2015). Accelerated deforestation in the humid tropics from the 1990s to the 2000s. *Geophysical Research Letters*, **42**, 3495–3501.

Kormos, R., Boesch, C., Bakarr, M. I., and Butynski, T. M. (2003). *West African Chimpanzees. Status Survey and Conservation Action Plan*. Gland: IUCN.

Manel, S., Williams, H. C., and Ormerod, S. J. (2001). Evaluating presence-absence models in ecology: the need to count for prevalence. *Journal of Appied Ecology*, **38**, 921–931.

McLennan, M. R. (2008). Beleaguered chimpanzees in the agricultural district of Hoima, western Uganda. *Primate Conservation*, **23**, 45–54.

Merow, C., Smith, M. J., and Silander, J. A. (2013). A practical guide to MaxEnt for modeling species' distributions: what it does, and why inputs and settings matter. *Ecography*, **36**, 1058–1069.

Mittermeier, R. A., Gil, P. R., Hoffman, M., *et al.* (2005). *Hotspots Revisited: Earth's Biologically Richest and Most Endangered Terrerestrial Ecoregions*. Mexico City: Cemex.

Myers, N., Mittermeier, R. A, Mittermeier, C. G., da Fonseca, G. A., and Kent, J. (2000). Biodiversity hotspots for conservation priorities. *Nature*, **403**, 853–858.

Nellemann, C. and Newton, A. (2002). The great apes: the road ahead. A Globio perspective on the impacts of infrastructural development on the great apes. See: www.globio.info/downloads/249/Great+Apes+-+The+Road+Ahead.pdf.

Palumbo, I., Rose, R. A., Headley, R. M. K., *et al.* (2017). Building capacity in remote sensing for conservation: present and future challenges. *Remote Sensing in Ecology and Conservation*, **3**, 21–29.

Petersen, R. and Pintea, L. (2017). Forest watcher brings data straight to environmental defenders. See www.wri.org/blog/2017/09/forest-watcher-brings-data-straight-environmental-defenders.

Pimm, S. and Askins, R. (1995). Forest losses predict bird extinctions in eastern North America. *Proceedings of the National Academy of Sciences of the United States of America*, **92**, 9343–9347.

Pimm, S. L. and Brooks, T. (2013). Conservation: forest fragments, facts, and fallacies. *Current Biology*, **23**, R1098–R1101.

Pimm, S. L., Jenkins, C. N., Abell, R., *et al.* (2014). The biodiversity of species and their rates of extinction, distribution, and protection. *Science*, **344**, (6187).

Pintea, L. (2005). Satellite analysis of threats to Gombe chimpanzees. In Caldecott J. and Miles, L., eds., *World Atlas of Great Apes and Their Conservation*. Cambridge: UNEP World Conservation Monitoring Centre, pp. 223–224.

Pintea, L. (2007). Applying satellite imagery and GIS for chimpanzee habitat change detection and conservation. PhD Thesis, University of Minnesota, MN.

Pintea, L. (2016). Geodesign restores chimpanzee habitats in Tanzania. *ArcNews*, Summer newsletter. See www.esri.com/esri-news/arcnews/summer16articles/geodesign-restores-chimpanzee-habitats-in-tanzania.

Pintea, L., Bauer, M. E., Bolstad, P. V, and Pusey, A. (2002). Matching multiscale remote sensing data to inter-disciplinary conservation needs: the case of chimpanzees in Western Tanzania. In *Pecora 15/Land Satellite Information IV/ISPRS Commission I/FIEOS 2002 Conference Proceedings*, p. 12.

Plumptre, A. J., Nixon, S., Vieilledent, G., *et al.* (2015). *Status of Grauer's Gorilla and Chimpanzees in Eastern Democratic Republic of Congo: Historical and Current Distribution and Abundance*. New York, NY: Wildlife Conservation Society.

Pusey, A. E., Pintea, L., Wilson, M. L., Kamenya, S., and Goodall, J. (2007). The contribution of long-term research at Gombe National Park to chimpanzee conservation. *Conservation Biology*, **21**, 623–634.

Reymondin, L., Jarvis, A., Perez-Uribe, A., *et al.* (2012). A methodology for near real-time monitoring of habitat change at continental scales using MODIS-NDVI and TRMM. See www.terra-i.org/dam/jcr:508a0e27-3c91-4022-93dd-81cf3fe31f42/Terra-i Method.pdf.

Saura, S. (2004). Effects of remote sensor spatial resolution and data aggregation on selected fragmentation indices. *Landscape Ecology*, **19**, 197–209.

Sexton, J. O., Song, X.-P., Feng, M., *et al.* (2013). Global, 30-m resolution continuous fields of tree cover: Landsat-based rescaling of MODIS vegetation continuous fields with lidar-based estimates of error. *International Journal of Digital Earth*, **6**, 427–448.

Stevens, S. S. (1946). On the theory of scales of measurement. *Science*, **103**, 677–680.

Stickler, C. M. and Southworth, J. (2008). Application of multi-scale spatial and spectral analysis for predicting primate occurrence and habitat associations in Kibale National Park, Uganda. *Remote Sensing of Environment*, **112**, 2170–2186.

Strassburg, B. B. N., Rodrigues, A. S. L., Gusti, M., *et al.* (2012). Impacts of incentives to reduce emissions from deforestation on global species extinctions. *Nature Climate Change*, **2**, 350–355.

Tracewski, Ł., Butchart, S. H. M., Donald, P. F., *et al.* (2016a). Patterns of twenty-first century forest loss across a global network of important sites for biodiversity. *Remote Sensing in Ecology and Conservation*, **2**, 37–44.

Tracewski, Ł., Butchart, S. H. M., Di Marco, M., *et al.* (2016b). Toward quantification of the impact of 21st-century deforestation on the extinction risk of terrestrial vertebrates. *Conservation Biology*, **30**, 1070–1079.

UNESCO (1973). *International Classification and Mapping of Vegetation, Series 6, Ecology and Conservation*. Paris: United Nations Educational, Scientific and Cultural Organization.

Vitousek, P. M. (1997). Human domination of Earth's ecosystems. *Science*, **277**, 494–499.

Wich, S. A., Garcia-Ulloa, J., Kühl, H. S., *et al.* (2014). Will oil palm's homecoming spell doom for Africa's great apes? *Current Biology*, **24**, 1659–1663.

Woodcock, C. E., Allen, R., Anderson, M., *et al.* (2008). Free access to Landsat imagery. *Science*, **320**, 1011.

Wrangham, R. W., Hagel, G., Leighton, M., *et al.* (2008). The Great Ape World Heritage Species Project. In Stoinski, T. S., Steklis, H. D., and Mehlman, P. T., eds, *Conservation in the 21st Century: Gorillas as a Case Study*. New York, NY: Springer Science and Business Media, pp. 282–296.

5 Wildfire Monitoring with Satellite Remote Sensing in Support of Conservation

Ilaria Palumbo, Abdoulkarim Samna,
Marco Clerici, and Antoine Royer

5.1 INTRODUCTION

Vegetation fires are common in many terrestrial ecosystems and fire has a key ecological role, being both a natural and anthropogenic component of most biomes in the world. Globally, between 3,500,000 and 4,500,000 km^2 burn every year. The largest areas are in the tropics, but a high occurrence of fires is also observed in the boreal forests of North America and Russia. Global statistics of fire activity based on satellite observations indicate the African continent to be the most affected, accounting for more than half of the total burned area (Tansey *et al.* 2004). In Africa, the average annual burned area is about 2,500,000 km^2 (Roy *et al.* 2008) compared to burned areas of 40,000 km^2 (Roy *et al.* 2008) in North America, 51,000 km^2 (Kasischke 1999, Zhang *et al.* 2003) in the Russian Federation, and 630,000 km^2 in Australia–Oceania (Roy *et al.* 2008). Fires caused by lightning are common in boreal forests but can also occur in the savanna, especially at the beginning of the dry season. However, in the tropics, the vast majority of burning is human-induced. Fires are often drivers of land-cover change and land degradation processes. They are also responsible for the release of large amounts of greenhouse gases (Kaufmann *et al.* 1996).

In this chapter, we discuss how satellite remote sensing can be used for fire monitoring and management in protected areas and in support of conservation action. We introduce basic remote sensing principles of fire detection and the current satellite instruments used for fire monitoring, including a discussion of current fire products and

119

the factors affecting their accuracy. The ecological role of fire in the major biomes is presented with explanations and examples about the main effects of fire on ecosystems. Finally, we provide several case studies about how information of fire activity derived from Earth observation is used to address conservation needs. The examples come from African institutions and protected areas that use two systems developed by European Commission projects. The first system, the eStation network, consists of a network of receiving stations installed in most sub-Saharan African countries. Each eStation receives daily Earth observation-based information that is used for environmental monitoring. The second platform is the Fire Monitoring Tool (FMT), a decision-support tool developed for conservation monitoring and management at the global level. The methods to derive information from satellite data are described, with particular attention to how protected areas and institutions benefit from these tools. Finally, we discuss the challenges associated with the use of satellite-derived information in conservation and how this situation can be improved upon in the future.

5.2   FIRE MONITORING WITH SATELLITE REMOTE SENSING

When available in near real time, information on fire occurrence from satellite systems can reduce the risk and damage caused by fire to human beings, goods, and ecosystems. The information required to study fire activity is: (i) fire risk, for prevention and early warning, (ii) real-time monitoring of fire occurrence, for prompt intervention and emergency management, and (iii) post-fire damage measurement, to understand and mitigate impacts. In some cases, it is possible to get this information directly from ground observations, for example through local patrolling that is provided by the fire brigades, the forest service, or local residents. However, over large or remote areas, ground-based information is often not available or difficult to obtain. Satellite remote sensing offers a unique means to observe fire activity on Earth, on a regular basis and at global level. This guarantees continuous monitoring, even in areas where there is no easy access on the ground. At present, information on fire occurrence from satellite systems is available in near real time, and several satellite instruments and missions (e.g. the Advanced Very High Resolution Radiometer (AVHRR), the Moderate Resolution Imaging Spectroradiometer (MODIS), and Landsat) have contributed to build long time-series of observations that can be used to study the history of fire over the past 30 years.

Fire Spectral Properties and Satellite Product Types

Fires have specific physical characteristics that allow their identification from spaceborne instruments. These features are associated with both the visible and the thermal spectral regions. Depending on the time of observation, we can identify or measure (i) fire fronts, smoke, temperature, and its intensity (fire radiative power) or (ii) the effects of combustion on vegetation or the land surface, after the fire passage. The first feature is observed while fire is occurring and it is called the *active fire* or *hotspot*, whereas the second feature is named the *burned area* or *burn scar*. Because of their strong spectral differences, these features are identified using different approaches. For further reading on fire detection from satellite instruments, see Roy *et al.* (2013).

Active fires show a thermal anomaly in the fire-affected pixels, generated by the contrast between the high temperature of one or more fire fronts and the surrounding environment. Therefore, they are typically observed from satellite sensors with infrared detectors; in particular, the most relevant spectral bands used for fire observation are the mid infrared (MIR 3.5–5 μm) and thermal infrared (TIRS 10–12 μm) (Chapter 2). The MIR is the spectral domain most sensitive to the temperature range of most fires, which release their peak of energy around 3.9 μm. Fire fronts and smoke can be observed using the visible bands, but it is not common practice to include them in algorithms for fire detection as these are unstable features to observe. As the active fire(s) is usually only present in a small fraction of the pixel, the minimum size of fire front(s) that can be detected depends on the pixel size, the sensor spectral bands (i.e. the sensor sensitivity to temperature), and the fire temperature. The higher the temperature, the smaller the portion of pixel that needs to be occupied by the active fire(s) to generate a detectable anomaly.

Satellite-derived active-fire products show the location and timing of burning, which are essential features to assess the diurnal fire cycle and fire seasonality. In addition, they also help to characterise the fire intensity by providing information on the instantaneous energy emitted by fire, also known as the Fire Radiative Power (or 'FRP').

These fire products represent only the fire activity observed at the time of the satellite overpass. Consequently, the lower the frequency of observation, the higher the percentage of missed fires. Another factor often limiting fire detection is cloud cover, particularly frequent in tropical humid regions. Combined with other sensor characteristics (e.g. low sensor spatial resolution), these factors can generate *omission errors*. By contrast, false detections can also occur, leading to *commission errors*. False detections are often caused by sun glare on smooth bright surfaces such as bare rocks, or other objects that generate a thermal anomaly (e.g. gas flares). Modern algorithms for fire detection have significantly reduced the number of false alarms by filtering known sources of heat, such as active volcanoes and gas flares (Schroeder *et al.* 2014).

Burned areas can be identified by the strong modification of the spectral properties of the surface after the fire passage. These changes are detected in the reflectance values in the visible and near/middle-infrared domains (0.4–3.5 μm) and the emissivity values in the TIRS domain (10–12 μm). The main changes in the surface properties are due to (i) a decrease in vegetation cover and water content (fuel load and fuel moisture content), which causes a decrease of reflectance in the near and shortwave infrared wavelengths (NIR and SWIR, respectively), and (ii) a deposition of combustion products, such as charcoal and ash. Usually, ash is dark and causes a decrease in reflectance in the visible domain; however, if combustion is almost complete, a light-coloured, highly reflective ash is produced, which can lead to an increase of reflectance rather than a decrease. Finally, burning causes (iii) an increase of exposed soil.

The alteration of surface characteristics depends on vegetation type and condition before the fire and the combustion efficiency (fire type, duration, and intensity) (Pereira *et al.* 1999, Roy *et al.* 2005). The spectral modification can be observed for short or long periods of time, depending on vegetation type. In savanna and grassland ecosystems, the altered reflectance values are usually observed for short periods, because the dispersal of ash and the fast regrowth of grass can generate a spectral signal similar to that of the pre-fire condition within 2 weeks from the fire event. In forest ecosystems, on the other hand, a decrease in reflectance can be observed for many years if fire affects the tree canopy (Gerard *et al.* 2003). If only the understorey is destroyed, the burn scar can remain undetected due to the green tree canopy above the surface. A common method used for burned-area mapping relies on multiple observations to identify the changes in the reflectance values over time. Moreover, since burned areas can be confused with other surfaces, for example water bodies and bare soil (Eva *et al.* 1998), classification methods are usually based on multiple bands rather than single-channel values.

Burned-area products provide information about the extent of the fire-affected surfaces and their temporal patterns. They are used to assess fire effects on ecosystems (i.e. fire severity) and are key inputs to biogeochemical models that are used to quantify gaseous emissions from biomass burning. Depending on the size of the pixel, a burned area can occupy a small portion or the complete extent of the pixel area. Therefore, coarse-spatial-resolution sensors tend to underestimate small burned areas as these occupy too small a fraction of their pixel area. However, high-spatial-resolution sensors (e.g. pixel size: 10–30-m) may have high *omission errors* due to their low revisit frequency (e.g. around 16 days for Landsat). Therefore, when considering the quality of satellite-based fire products, it is important to evaluate both the spatial and temporal resolution of the sensor. And, as a general recommendation, fire activity can be better assessed using active-fire and burned-area information combined, as they provide complementary data.

Overview of Satellite Sensors and Derived Fire Products

Satellite remote sensing has been used for environmental monitoring since the 1970s, but it was only in the 1990s that global fire activity was first assessed using satellite Earth observation. The first maps of fire occurrence were derived from the AVHRR sensor (Giglio *et al.* 1999). These maps showed the spatial distribution of active fire in the global biomes and their temporal patterns (Dwyer *et al.* 2000). These were important insights for the time, and they highlighted how fire encompasses a variety of biomes around the world. However, the first sensors used for fire monitoring were not specifically designed for this purpose and presented a number of limitations. Being mainly developed for research in meteorology, climate, oceanography, and vegetation dynamics, these sensors did not measure at the optimum spectral wavelengths for fire detection or mapping. So, for example, the large amount of energy released by intense fires could cause sensor saturation, which prevented fire detection; or the sensors did not have sufficient spatial resolution to observe most burn scars.

Around the year 2000, sensors could measure thermal and optical spectral ranges that are more sensitive to fire, and had higher spatial resolutions, which improved the overall accuracy of detection. As satellite data became available for longer time-series, the first multi-annual fire products were released. An example is the World Fire Atlas (1995–2012), based on night-time observations of active fires acquired by the Advanced Thermal Scanning Radiometer (ATSR) instrument. Among the burned-area products, one of the first multi-annual maps released was the L3JRC burned area map derived from Satellite Pour l'Observation de la Terre (SPOT) Vegetation imagery (2000–2007).

MODIS was the first sensor with spectral bands there were especially designed for fire detection (Chapter 2). The same sensor is on two satellites, and thus provides four daily observations (day and night), which allow the overall representation of the diurnal fire cycle.

MODIS is currently the most widely used satellite sensor for mapping both active fires and burned areas. MODIS fire products are available at 500-m and 1-km spatial resolution and can be used for regional and global monitoring. As a MODIS mission follow-up, the Visible Infrared Imaging Radiometer Suite (VIIRS) has higher spatial resolution (750-m and 375-m) and improved detection of small fires. The fire detection algorithm has been further improved so that VIIRS-based fire products have better accuracy (Schroeder *et al.* 2014). The 375-m spatial-resolution products are distributed by the National Aeronautics and Space Administration (NASA) Fire Information for Resource Management System (FIRMS, Davies *et al.* 2009), while the University of Maryland distributes the product with 750-m spatial resolution (http://viirsfire.geog.umd.edu). More information about MODIS and VIIRS fire products is available at https://earthdata.nasa.gov/earth-observation-data/near-real-time/firms. New fire products have been also released by the European Space Programme Copernicus (www .copernicus.eu), such as the global burned-area map based on Proba-V imagery (http://land.copernicus.eu/global/products/ba).

Multi-annual fire products have highlighted fire seasonality and spatial patterns, including the inter-annual variability, often due to weather and climatic conditions, or their anomalies. These observations have shown the global relevance and impact of biomass burning on ecosystem ecology, climate, and human livelihoods. Geostationary systems can also be used to monitor fires (see http://wfabba.ssec.wisc .edu and www.eumetsat.int for more information).

A summary of past and current fire products derived from satellite sensors is provided in Table 5.1. It reports most of the operational global products and some regional applications, but is not an exhaustive list.

Factors Affecting the Accuracy of Satellite-Based Fire Products

The key technical features affecting the sensors' observation capabilities are their temporal, spatial, and radiometric resolution (Chapter 2). The temporal resolution is defined by the frequency of observation.

Table 5.1 *Fire products and their main characteristics*

Product type (and name where relevant)	Satellite sensor	Period covered	Temporal frequency	Extent	Spatial resolution	References and notes
Active fire	ATSR/AATSR (on-board ERS-1, ERS-2 and Envisat satellites)	ATSR 1995–2002 AATSR 2003– 2012	Monthly	Global	1-km	Schultz 2002; http://due.esrin .esa.int/page_ wfa.php
Burned area, Global Burned Area (GBA 2000)	Vegetation (on-board SPOT series)	2000	Daily	Global	1-km	Grégoire *et al.* 2003; http://for obs.jrc.ec .europa.eu/pro ducts/burnt_ar eas_gba2000/gl obal2000.php
Burned area, L3JRC	Vegetation (on-board SPOT series)	2000–2007	Daily	Global	1-km	Tansey *et al.* 2008; http://for obs.jrc.ec

Table 5.1 (*cont.*)

Product type (and name where relevant)	Satellite sensor	Period covered	Temporal frequency	Extent	Spatial resolution	References and notes
						.europa.eu/products/burnt_areas_L3JRC/GlobalBurntAreas2000-2007.php
Burned area, MCD45	MODIS (Terra and Aqua satellites)	2002 – ongoing	Monthly	Global	500-m	Roy *et al.* 2005; http://modis-fire.umd.edu/pages/BurnedArea.php?target=Download
Active fire, MDL14DL	MODIS (Terra and Aqua satellites)	2002 – ongoing	Daily	Global	1-km	Giglio *et al.* 2003a; https://earthdata.nasa.gov/earth-observation-data/near-real-

Table 5.1 (*cont.*)

Product type (and name where relevant)	Satellite sensor	Period covered	Temporal frequency	Extent	Spatial resolution	References and notes
						time/firms/c6-mcd14dl
Burned area	AVHRR-GAC (Global Area Coverage) on-board NOAA series	1981–1983, 1985–1991	Annual	African continent	5-km	Barbosa *et al.*, 1999; http://forobs.jrc.ec.europa.eu/products/burnt_areas_africa_81-91/africa81-91.php
Active fire	AVHRR (on-board NOAA series)	1992-1993	Monthly	Global	1-km	Dwyer, *et al.*, 1999, 2000; http://forobs.jrc.ec.europa.eu/products/fire_occuren ce_gfp_92-93/gl

Table 5.1 (*cont.*)

Product type (and name where relevant)	Satellite sensor	Period covered	Temporal frequency	Extent	Spatial resolution	References and notes
						obal-fire-product1992-93 .php
Active fire	VIIRS (on-board Suomi National Polar-orbiting Partnership, SNPP)	2012 - ongoing	Daily	Global	375-m / 750-m	Schroeder *et al.* 2014; https://ea rthdata.nasa .gov/earth-observation-data/near-real-time/firms/viir s-i-band-active-fire-data
Active fire	VIRS (on-board the Tropical Rainfall Measuring Mission, TRMM)	1998 - 2015	Monthly	Tropics	0.5°	Giglio *et al.*, 2003b. https:// disc.gsfc.nasa .gov/precipita tion/addi tional/tools/tr mmVirsFire .html

Table 5.1 (cont.)

Product type (and name where relevant)	Satellite sensor	Period covered	Temporal frequency	Extent	Spatial resolution	References and notes
Burned area	Landsat series	Ad-hoc products, since the 1970s	Several months	Regional to local studies	30-m	Boschetti et al. 2015.
Burned area	Advanced Wide Field Sensor (AWiFS) (on-board Indian Remote Sensing, IRS)	2014 (fire season)	Monthly	Regional	56-m	Sudhakar Reddy et al. 2017.
Burned area	Vegetation (on-board Proba-V)	2014- ongoing	Every 10 days	Global	300-m	Pre-operational stage; http://land.copernicus.eu/global/products/ba

More than one observation per day is required to assess the diurnal fire cycle, based on active-fire occurrence. However, high-frequency image acquisition has been, so far, mainly provided by low-spatial-resolution sensors, which tend to miss small fires (e.g. fire size less than 100 m^2) (Giglio *et al.* 2003a). The effect of spatial resolution on the fire detection performance is discussed in Schroeder *et al.* (2014). In general, a higher spatial resolution increases the chance of detection of small fires, although very high resolution (<10-m) can also result in problems of fire detection due to signal saturation. The spatial, temporal, and spectral resolutions of sensors are often interlinked, and the optimal conditions for fire detection and burned-area mapping are a compromise between them.

The accuracy in burned-area estimates varies greatly across world biomes. The main cause of uncertainty in the estimates is the low spatial resolution of the data. For example, MODIS only detects burned areas larger than around 100–120 ha. Small fires represent around 35 per cent of global burned areas and are therefore important for continental- and global-scale environmental-change assessment and climate-change analysis. Understorey burned areas can also be missed; higher spatial resolution would improve their observation in dense forests if accompanied by increased temporal resolution (one or more observation per week). Frequent observations are needed because fires are very dynamic, often fast moving, and fire-affected areas may change rapidly after the fire passage, or be obscured by clouds (e.g. in the tropics).

Medium-resolution satellite sensors, such as Landsat 5 and subsequent Landsat missions (30-m), have been used to produce a finer characterisation of the burned areas, but their 16-day revisiting time is usually too low to provide a consistent observation of these areas, causing large omission errors (Boschetti *et al.* 2015). Frequent observations are particularly important for accurate burned-area mapping in savanna and grassland ecosystems, where removal of ash and fast vegetation regrowth can mask out the effect of fire within a few weeks after the fire event. The demand for more accurate burned-area products over longer time-series is increasing, not only among climate and vegetation modellers but also ecologists and land managers who are dealing with regional or landscape assessment and planning.

The new Sentinel missions are expected to improve the current fire monitoring capacity and the accuracy of fire maps. Verheggen *et al.* (2016) combined Sentinel-1A (Synthetic Aperture Radar (SAR)) and Sentinel-2A (MultiSpectral Instrument (MSI)) data to map burned areas in the Republic of Congo. This combination overcame some of the instruments' limitations for an improved identification of burned areas. The Sentinel-1A radar system (available every 15 days with 25-m spatial resolution) allowed observation of fire-affected areas even in the presence of clouds and haze. Both are very common in tropical regions and often prevent Earth observation using optical instruments. Sentinel-2A data were used to produce an independent burned-area map of the region; these data have a spatial resolution similar to those from Landsat, but with a shorter revisit time (10 days). Both the Sentinel-1 and -2 systems now have a twin sensor to work as a constellation, further increasing their frequency of observation.

5.3   FIRE IN THE GLOBAL BIOMES AND CONSERVATION CHALLENGES

Fire plays an important role in the global biogeochemical cycles and its impact on climate processes is widely recognised (Li *et al.* 2014). At regional and local levels, fire often represents a hazard to human health and safety, is a major cause of air pollution (e.g. peatland fires in Southeast Asia; Page and Hooijer 2016), and is a threat to natural resources and biodiversity. On the other hand, fire played an important ecological role in the evolutionary and ecological history of many species and biomes (Bowman *et al.* 2013) and continues to be key to species survival and ecosystem functions. Therefore, its positive and negative effects need to be deeply understood to make conservation efforts effective in the short and long term.

Fire Ecology

The Nature Conservancy's Global Fire Initiative (Shlisky *et al.* 2007) is one of the most comprehensive studies on fire's effects on ecosystems. Ecosystems are divided into three main categories on the basis of their adaptation and response to fire: fire-independent, fire-sensitive, and fire-dependent (Figure 5.1). An additional class, the fire-influenced ecosystems, is also identified as a transitional category between the three classes. The ecosystem categories are described in Box 5.1. The classification considers intact ecosystems, where fires occur within the range of frequency and intensity to which ecosystems are adapted. However, if natural fire regimes (i.e. fire seasonality, frequency, etc.) are modified, the overall fire-activity changes, often with negative effects on species richness, composition, and structure, and this can lead to biodiversity loss. Fire regimes change according to ecosystems, weather, and climate, but they are also modified by human activities (Hardesty *et al.* 2005, Myers 2006).

What is a Fire Regime?

A fire regime is a combination of the dominant temporal, spatial, and behaviour patterns and the type of fire in a certain biome or a region. The temporal patterns are characterised by the frequency and seasonality of fire. The frequency refers to the duration of time between two consecutive fires in the same area, whereas the seasonality describes the period of the year when most fire activity usually occurs (i.e. the fire season). The spatial patterns are related to the location and extent of the fire-affected areas during the fire season (e.g. the number of fires and the average surface burned). The behaviour patterns include the flame height, the spread rate, and the fire intensity and severity. The fire severity describes the effects of fire on the ecosystem, the degree of mortality, and the effects on the soil. The fire type is related to the part of vegetation structure most affected by burning: so there can be ground fire, which affects the organic material on the ground;

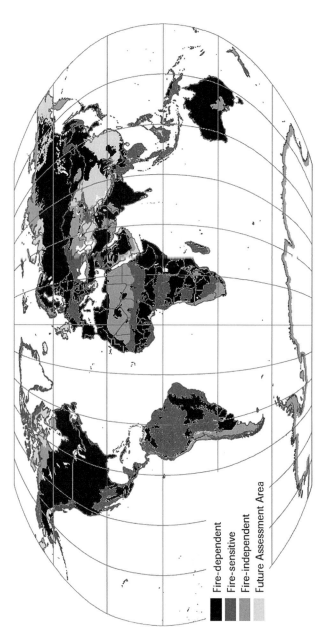

FIGURE 5.1 The distribution of fire-dependent, fire-sensitive, and fire-independent ecosystems in the world (from Shlisky *et al.* 2007).

Fire-dependent
Fire-sensitive
Fire-independent
Future Assessment Area

BOX 5.1   **Ecosystem types according to their response to fire (from Shlisky *et al.* 2007)**

Fire-independent ecosystems – around 15 per cent of the global terrestrial area.

Fire-independent ecosystems do not experience fires, mainly because the climatic and environmental conditions prevent fire ignition or its spread. These ecosystems are found in regions with cold or very humid climates or where fuel is almost absent (e.g. deserts and tundra biomes).

Fire-sensitive ecosystems – around 22 per cent of the global terrestrial area.

Fire-sensitive ecosystems have not evolved in the presence of fire and, in healthy conditions, they are rarely exposed to burning. They are located in wet or cool climate regions and their vegetation type and structure tend to hinder fire ignition or spread. However, when natural conditions are altered, they can become vulnerable to fire and be negatively affected. Examples of these ecosystems are found in the tropical moist broadleaf forests of the Amazon, in Southeast Asia, and in the Congo Basin. In these forests, deforestation has modified the hydrological balance and vegetation structure, making ecosystems much more vulnerable to fire than before. The increased fire activity can start a vegetation conversion process that favours fire-prone species and causes more frequent fires to occur.

Fire-dependent ecosystems – around 53 per cent of the global terrestrial area.

In fire-dependent ecosystems, most species have been exposed to fire during their evolution, so they have developed adaptations to quickly recover or take advantage of the burning. In such ecosystems, such as the savanna and the coniferous forests of tropical and boreal regions, fire plays a substantial role in maintaining species structure, composition, and biodiversity (Komarek 1983, Archer 2000, Laris 2011).

Fire-influenced ecosystems

Fire-influenced ecosystems lie in the transition zone between fire-sensitive and fire-dependent or fire-independent ecosystems. They are sensitive to fire but can include species with some adaptations to fire, so that fire may play a role in creating or maintaining some niches or habitats.

surface fire, affecting the lowest vegetation layer; and crown fire, which burns the upper part of the tree or shrub canopy.

The modification of one or more of the fire-regime components affects the others and alters the overall fire regime. Only 25 per cent of the terrestrial ecoregions exhibit intact (natural) fire regimes, over 50 per cent have degraded fire regimes, and 8 per cent show very degraded conditions (Shlisky *et al.* 2007). The most natural fire regimes are found in the boreal forests and taiga, whereas Mediterranean forests, woodlands and scrub, and broadleaf forests have the most degraded fire regimes. The modification of ecologically appropriate fire regimes, more than fire itself, is identified as a source of threat to ecosystems and biodiversity, because altered fire regimes affect fuel availability and its spatial distribution, with effects on ecosystem structure and functioning.

Traditional and Modern Use of Fire

In many regions of the world, fire has been used since prehistoric times and still today as a tool to modify landscapes and access natural resources (Scholes and Archer 1997).

When used for land management, fire is ignited under controlled conditions to achieve the desired effects; therefore, the burning conditions and timing are carefully planned. For example, in the African and Australian savannah, controlled burning is used for fuel management in order to protect settlements and infrastructures.

Fuel management practices that involve controlled burning have proved to be effective and beneficial to the landscape; however, in the twentieth century, many national governments (e.g. Canada, United States, Australia) adopted fire exclusion and suppression policies to protect infrastructures, properties, and natural resources. But these policies increased the build-up of flammable vegetation, with the consequent increase of fire risk and severity, especially in fire-prone regions. In these regions, the fire-suppression policies had detrimental effects on health and safety, and they also affected biodiversity and natural resources.

Fire Management for Conservation

As a land management tool, fire can be used under controlled conditions to achieve positive effects on habitats and ecosystems, and often protects them from unwanted subsequent fires. Managed fires are used to promote biodiversity and increase landscape variability, to maintain some key niches in the ecosystems. However, in many conservation areas, fires can be a threat to habitats and biodiversity, and are often associated with illegal activities, such as poaching. Suppressing fires can be problematic: fire exclusion in protected areas with fire-adapted ecosystems can cause negative effects on habitats, biodiversity, and livelihood (Arno 1996). Learning from this experience, many protected areas have restored natural fires through controlled burns, as in the Yellowstone National Park in the USA (Turner *et al.* 2003).

In tropical protected areas with savanna ecosystems (e.g. Australia and Africa), prescribed burning, e.g. by the creation of landscape mosaics with burned and unburned patches, is used to increase habitat variability and biodiversity. Fire is also applied to regenerate palatable grass for herbivores during the dry season, to control bush encroachment, and to prevent the spread of invasive species (Scholes and Archer 1997). Fire can thus facilitate wildlife protection from illegal hunting, as the availability of forage and resources will keep most wildlife within the conservation area, where they are better protected.

The preparation of fire plans improves preparedness and early warning, including fire suppression, where required. It also allows easier identification of threats and pressures associated with fire. For example, fires caused by poachers or by other illegal activities can be clearly distinguished from known management burnings.

Therefore, the successful implementation of conservation programmes also depends on a correct understanding of the role of fire in the ecological processes, and the application of an appropriate fire regime. Data are needed to provide an ecological assessment of the conservation area and its response to fire.

Fire-plan implementation is challenging because protected areas are often very large and can be too remote to be regularly surveyed and patrolled. Human resources and costs can be high because specialised personnel may be required and specific safety equipment is needed. Fire plans also require verification, through patrolling, monitoring, and surveying activities over time, in order to guarantee that objectives are achieved, and habitats and wildlife are protected.

Many of these challenges can be addressed with satellite remote sensing, as this offers a means for habitat monitoring over large and remote areas, and over long periods.

5.4 SATELLITE REMOTE SENSING-BASED TOOLS FOR FIRE MANAGEMENT

In this section, we describe two systems that distribute information from satellite remote sensing, to support environmental monitoring and nature conservation. These systems are the outcome of projects that aim to increase capacity in information management and the use of Earth observation for monitoring, analysis, and decision-making in conservation.

The eStation system (http://estation.jrc.ec.europa.eu) was developed at the Joint Research Centre (JRC) of the European Commission during the long-term project, Monitoring for Environment and Security in Africa (MESA). MESA follows up the activities of projects that started in 2001, in the context of the Global Monitoring for Environment and Security (GMES) and Africa initiative. This project was aimed at improving the sustainable management of natural resources, biodiversity conservation, and food security at continental, regional, and national levels. A network of satellite receiving stations (the eStations) was established in Africa, in 2012, to supply Earth observation information to most sub-Saharan African countries, in particular serving the Ministries of Agriculture, Natural Resources, and the Environment, and research institutions dealing with natural resource management and conservation programmes. This community receives user-oriented products, including environmental indicators of wildfires, derived from satellite observation on a regular basis.

The second system is the web-based dashboard for fire monitoring, FMT, which was developed at JRC to serve MESA and the Biodiversity and Protected Areas Management (BIOPAMA) Programme (http://www.biopama.org/). The BIOPAMA programme was started in 2012 as a joint initiative between the International Union for the Conservation of Nature (IUCN) and the European Commission–JRC. Its mission is to provide scientific knowledge to support regional and national protected area management, and to improve capacity for policy- and decision-making in biodiversity conservation in African, Caribbean, and Pacific (ACP) countries. The FMT provides BIOPAMA and MESA users with information on fire occurrence and ecological indicators for protected area management and biodiversity conservation, with the aim of increasing management effectiveness and governance.

Although they were developed for different projects, the eStation network and FMT share some technology and datasets. In particular, the eStation server hosted at JRC is used for some of the FMT data acquisition and processing, while the FMT provides the eStation network with some of the web services that are used to produce the fire indicators distributed through its network.

The eStation Network

The eStation network is composed of satellite receiving stations installed in 47 countries in sub-Saharan Africa (Clerici *et al.* 2013). It reaches a wide community of registered users, including the Ministries of Agriculture, Forestry, and Environment and regional institutions in the Regional Economic Communities (RECs) of the African Union (www.un.org/en/africa/osaa/peace/recs.shtml), (e.g. the Climate Prediction and Applications Centre (ICPAC) in eastern Africa).

Each eStation is a processing and visualisation server, which receives Earth observation-based products through the European

Organisation for the Exploitation of Meterorological Satellites (EUMETSAT) data broadcasting system (EUMETCast) and from other data sources available over the internet (e.g. NASA ftp servers). The system handles large volumes of data and runs the routines required for data acquisition, processing, and analysis. It then uses these data to generate environmental variables and indicators. All the processing steps are predefined and scheduled on each eStation to avoid manual data manipulation. This allows users to generate their own environmental analysis and reports without having to deal with heavy processing and data management. Advanced users can modify the system and customise it, according to their needs. The eStation includes a user interface that allows data display, map and chart production, the analysis of time-series of environmental variables, and the generation of reports and bulletins.

Data received at the eStations cover a wide range of thematic areas, such as natural resource management, wildfires, habitat conservation, forestry and rangeland management, land degradation mitigation, and coastal and marine resources management. The system also includes climate services for disaster risk reduction and food security. Environmental monitoring is possible at regional and national levels; however, information distributed through the eStation is also available for local administrative units and protected areas. Information is based on a multitude of Earth observation systems: SPOT–Vegetation, Proba-V–Vegetation, Meteosat Second Generation (MSG)–Spinning Enhanced Visible and InfraRed Imager (SEVIRI), and MODIS–Terra/Aqua.

A major added value of this system is its independence from broadband availability, since data delivery is mostly done through the EUMETCast broadcasting system. This is particularly critical in African countries, where internet connection can be limited.

The FMT: A Decision-Support System

The FMT was developed at JRC to give free access to near-real-time and historical information about global fire activity, which is derived from satellite observations. The tool is a publicly accessible web portal (http://firetool.jrc.ec.europa.eu). The system was designed to address the needs of a wide user community that is working in conservation and land management, including protected-area managers, conservation institutes, and ministries. The tool supports the prevention, planning, and control of fire in conservation areas and can be used for ecological monitoring and habitat assessment at a site level, and also regionally. Because users include professionals with different expertise and needs (e.g. ecologists, park managers, decision-makers), the tool was designed to respond to a range of requirements and various levels of complexity.

The FMT aims to facilitate the use of satellite-based information by professionals with no specific remote sensing background. The tool supplies spatially explicit information of fire occurrence together with ecological indicators that describe the fire regime and the current fire season, so it does not require any image processing or specific remote sensing knowledge. Moreover, the user can query the system to retrieve ecological metrics over specific periods and areas of interest. The FMT provides daily updates on the burning patterns inside and around the protected areas, ecological indicators, downloadable maps, and statistics. A complete description of the metrics and indicators is provided in Table 5.2. The information can be used for decision-making without the need of additional analysis, but users can also export data as geotiff and shapefile, if required, to use them in a local geographic information system (GIS) environment.

Fire monitoring is provided for the conservation sites listed in the World Database on Protected Areas (WDPA) of the IUCN World Conservation Monitoring Centre (WCMC) (UNEP 2010) and a buffer zone around each protected area. In addition, the tool includes other sites of interest, such as forest reserves and forest concessions. A graphical description of the system architecture is shown in Figure 5.2.

The information on fire activity is derived from MODIS fire products (i.e. active fire and burned area): the active-fire product (MCD14DL, Giglio *et al.* 2003a) is retrieved from the NASA FIRMS

Table 5.2 *Ecological metrics and indicators provided by the FMT*

Metrics and indicators	Input data	Description
Fire seasonality	Active fire	Timing of the fire activity, peak, and duration
Fire occurrence	Active fire	Fire count in the protected area and the buffer zone
Fire density	Active fire	Fire count by unit area
Fire intensity	Active fire	Fire radiative power, the energy emitted by the fire
Reference fire occurrence	Active fire	Averaged values of fire count based on the time-series (2003–2012)
Land-cover distribution	Land-cover data (Globcover 2005; htt p://due.esrin.esa.int/pa ge_globcover.php)	The vegetation classes distribution in the area of interest (as percentage)
Burned area extent	Burned area	The surface burned, in hectares and as a percentage of the park total area
Fire-affected vegetation	Burned area	The proportion of area burned in the vegetation classes
Reference burned area	Burned area	Averaged values of burned area based on the time-series (2003–2012)

(Davies *et al.* 2009), while the burned-area data are derived from the University of Maryland (MCD45, Roy *et al.* 2008). The active-fire product contains daily information of fire occurrence (time, location of burning, fire radiative power), based on four MODIS observations

FIGURE 1.2 An image of the Earth, as seen by the Apollo 17 crew.
(Image credit: NASA.) (A black and white version of this figure will appear
in some formats.)

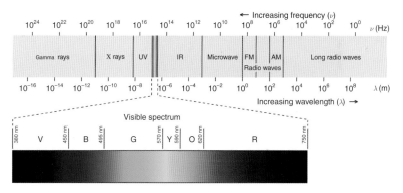

FIGURE 2.2 The electromagnetic spectrum. In the visible region, V =
violet, B = blue, G = green, Y = yellow, O = orange, R = red.
(Philip Gringer, Wikimedia Commons.) (A black and white version of this
figure will appear in some formats.)

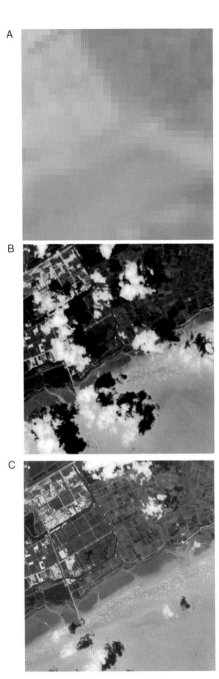

FIGURE 2.6 Different spatial resolutions resulting in different levels of detail with (A) the Terra/Aqua MODIS (300-m), (B) the Landsat 8 OLI (30-m), and (C) the Sentinel-2 MSI (10-m) sensors. Images come from different days in same week.

(A black and white version of this figure will appear in some formats.)

FIGURE 3.1 Reference photographs of tidal flats and typical threats in the Yellow Sea region of East Asia. (A) Tidal flats with extensive dendritic drainage networks with wind chop at the waterline, Gomso Bay, South Korea; note the loss to reclamation for aquaculture. (B) Large, low-sloping tidal flats in Jiangsu Province, China. (C) Coastal reclamation works being implemented in Dandong, Liaoning Province, China. (D) Tidal flats reclaimed for agriculture, urban development, and airport land, Incheon, South Korea.
Adapted from Murray *et al.* 2015a. (A black and white version of this figure will appear in some formats.)

FIGURE 8.1 (A) A blue whale surfacing near the shipping channel off southern California.
(A black and white version of this figure will appear in some formats.)

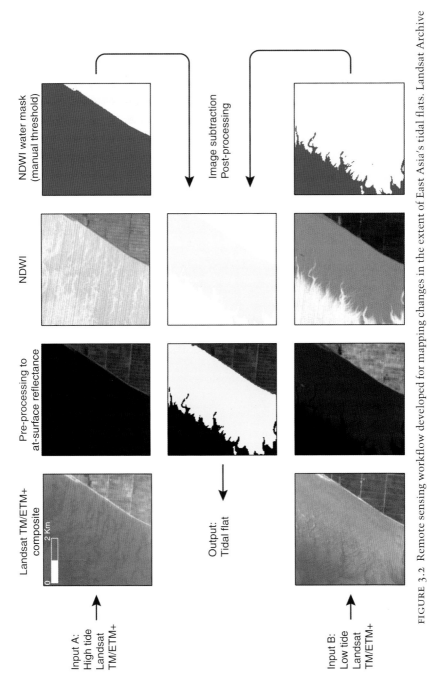

FIGURE 3.2 Remote sensing workflow developed for mapping changes in the extent of East Asia's tidal flats. Landsat Archive imagery is ingested, pre-processed, and classified into a two-class land–water image, which is then differenced among images acquired at high and low tide.

Adapted from Murray *et al.* 2012. (A black and white version of this figure will appear in some formats.)

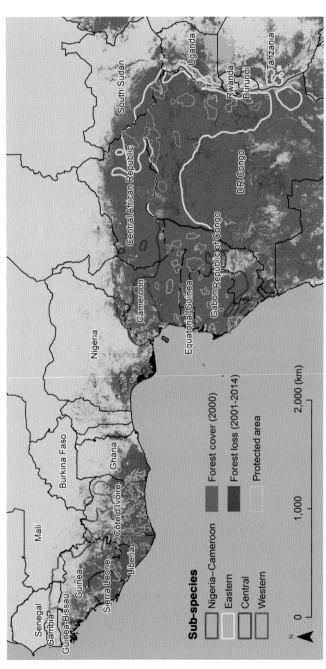

FIGURE 4.2 Forest extent (green) and loss (red) between 2001 and 2014 across the four chimpanzee ranges, with protected areas from WDPA outlined in light blue.
(A black and white version of this figure will appear in some formats.)

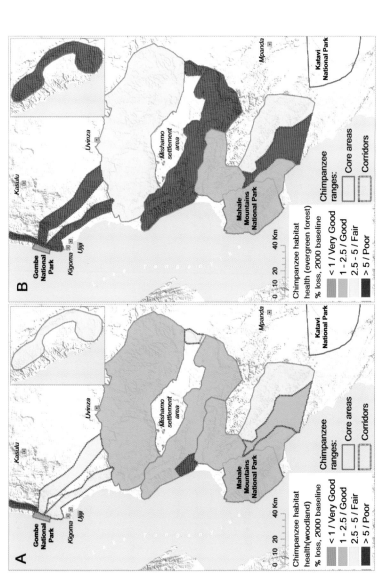

FIGURE 4.6 The status of chimpanzee habitat across two KEAs by management polygons, defined by the the chimpanzee core areas and corridors. (A) A size-category KEA, measured as per cent forest and woodland loss within a suitable habitat. (B) A condition KEA, measured as per cent evergreen forest loss within a suitable habitat. Conservation action planning stakeholders agreed to interpret chimpanzee habitat health as Very Good (dark green) if per cent forest loss was less than 1 per cent, Good (light green) as 1–2.5 per cent, Fair as 2.5–5.0 per cent and Poor if more than 5 per cent was lost compared to the 2000 baseline. (A black and white version of this figure will appear in some formats.)

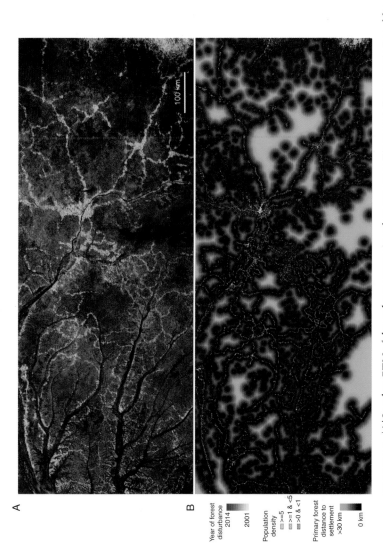

FIGURE 4.7 (A) Landsat ETM+ false colour composite of an area near Kisangani, Democratic Republic of the Congo. (B) A map of the same geographic area shows the relative distance from primary forest to human settlements (brighter green areas are located farthest away from human settlements), forest disturbance from 2001 to 2014 shown in a gradient from yellow to red, and population density shown in cyan tones.

(A black and white version of this figure will appear in some formats.)

Year of forest disturbance
2014
2001

Population density
>=5
>=1 & <5
>0 & <1

Primary forest distance to settlement
>30 km
0 km

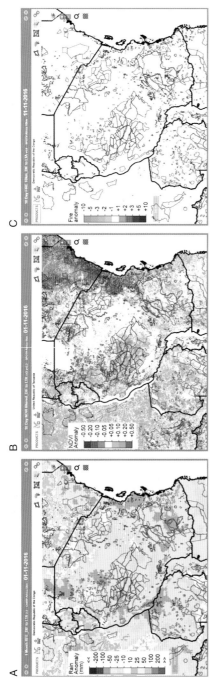

FIGURE 5.4 Example maps for Tanzania and its surrounding countries, derived from the eStation user interface. They are used for the continental assessment of wildfire, rainfall, and vegetation anomalies. The maps refer to the period 1–11 November 2016, with the WDPA boundaries displayed in dark grey. (A) The rain anomaly as derived from FWES RFE (source: NOAA's Climate Prediction Center; Xie *et al.*, 1997). (B) The NDVI anomaly calculated as the current NDVI/average NDVI (1999–2014). (C) The fire anomaly (on a 10-km grid) based on the MODIS active-fire product (1-km). (A black and white version of this figure will appear in some formats.)

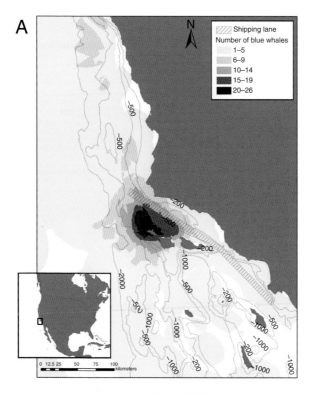

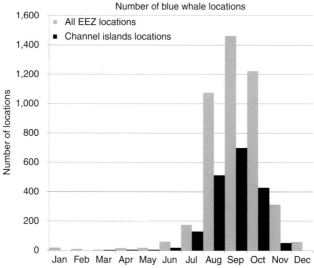

FIGURE 8.4 Number of overlapping blue whale core areas near shipping lanes off southern California (A) and central California (B). The inset maps show the location off the US west coast. Graphs show the number of blue whale locations in US west coast waters (grey) and in the area shown in the maps (black).

Adapted from Irvine *et al.* 2014. (A black and white version of this figure will appear in some formats.)

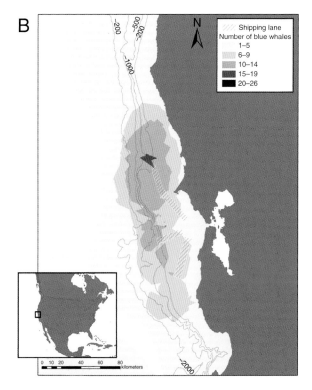

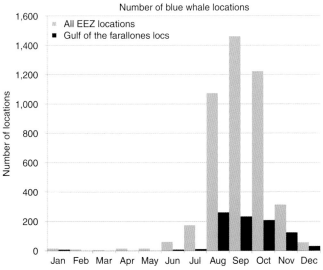

FIGURE 8.4 (*Continued*)

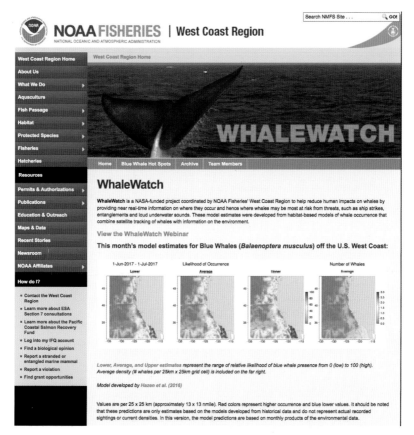

FIGURE 8.5 WhaleWatch webpage hosted on the NOAA/NMFS West Coast Region website (www.westcoast.fisheries.noaa.gov/whalewatch, accessed 6 July 2017). (A black and white version of this figure will appear in some formats.)

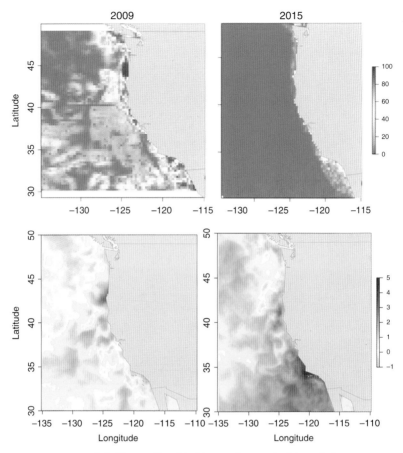

FIGURE 8.6 The predicted habitat preference for blue whales
in September 2009 (an average year for ocean conditions)
and September 2015 (an unusually warm period) off the US west coast
(upper panels, percent likelihood of occurrence) and the corresponding
satellite-derived SST anomaly for those periods (lower panels, °C).
Adapted from Hazen *et al.* 2016. (A black and white version of this figure
will appear in some formats.)

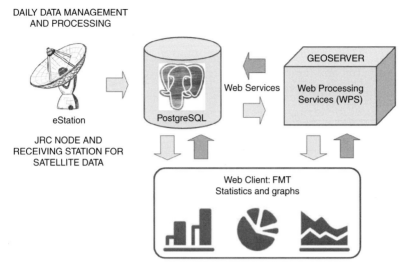

FIGURE 5.2 The FMT architecture.

per day. The product is distributed 48 hours after the satellite overpass, with 1-km spatial resolution. The burned-area product has a 500-m spatial resolution and informs on the extent of the fire-affected area and the approximate date of burn. It is distributed about 2 months after data acquisition because of processing constraints. Both fire products are available at a global level and cover the past 15 years.

The user defines the time period and geographic area for which the statistics of fire activity are produced. The statistics of a selected fire season are also compared with the averaged values of a 10-year time-series (2003–2012) to identify possible anomalies. The information can be used by park ecologists and managers to assess fire regimes in near real time, and over long periods of time. In particular, near-real-time information of fire occurrence is needed for: (i) patrolling activities, and the verification of the fire plans, if they exist; (ii) the identification of threats or pressures; (iii) a fast response to unwanted fires, due to human activities occurring in the protected-area vicinity or illegal activities carried out inside the park boundaries (e.g. poaching); and (iv) prioritisation of resources, through the identification of critical areas within the protected area. Historical data from time-series analysis are used to assess a fire regime in a protected area and its possible alteration, to identify anomalies in the spatial and temporal fire patterns, to identify persistent stress or pressure factors on the ecosystem, and finally, to assess the long-term effects of fire management plans and management effectiveness.

5.5   CASE STUDIES IN AFRICAN PROTECTED AREAS

Protected Areas and Government Institutions in Tanzania

The Wildlife Division and the Forest Department of the Ministry of Natural Resources and Tourism in Tanzania receive regular information on fire, environment, and climate from the eStation installed in their headquarters. These institutions work in close collaboration and use the eStation datasets to assess protected-area state and threats at national and regional levels. In this case study, we consider Selous Game Reserve, which, with an area of about 50,000 km², is one of the largest protected areas in Africa. Selous hosts a high variety of habitats, including miombo woodlands, open grasslands, riverine forests and swamps, and a rich and diverse wildlife, in particular, large populations of African elephant (*Loxodonta africana*), black rhinoceros (*Diceros bicornis*), African wild dog (*Lycaon pictus*), hippopotamus (*Hippopotamus amphibius*), and Cape buffalo (*Syncerus caffer*). This exceptional biodiversity makes Selous an outstanding site for conservation of biological diversity. However, its large size requires substantial resources for staff and management, and makes ecological monitoring more challenging, because of resource availability or accessibility issues in some remote areas.

Satellite information provided by the eStation allows monitoring of the park on a regular basis. Moreover, the environmental indicators provided by the eStation can be used for the assessment of the ecosystems and conservation areas around Selous. The Selous management plan includes prescribed burns, which are applied every year to selected plots. This practice creates areas with fresh grass, to ensure availability of pastures to wild animals throughout the dry season. The example shown in Figure 5.3 demonstrates how fire patterns and occurrence can be monitored from the eStation user interface. The information provided allows the assessment of the protected area in the context of other conservation sites, and supports the implementation of national and regional conservation policies.

In Figure 5.3, the map shows Selous Game Reserve in the context of other neighbouring WDPA protected areas. In this example, fire count was derived from the MODIS active-fire product and computed over a 10-km regular grid for the periods shown. The 10-km grid, rather than the original 1-km spatial resolution, is used to facilitate analysis at the national and regional scale. For the same reason, information on fire count is delivered with a 10-day time step instead of the original daily frequency of the MODIS active-fire data. Decadal information is also preferred to align with the frequency of other variables used in the analysis, such as rainfall and the vegetation index (Normalised Difference Vegetation Index, NDVI), which is a measure of the vegetation photosynthetic activity and status (for more information see Chapter 2). The bar chart shows fire activity (current and average active fire count) and the NDVI values (current and average values) on a 10-day basis. NDVI is used to study the vegetation phenology (i.e. the growing patterns from green up to senescence) and users can observe the distribution of fire count against the vegetation indicator. Fire activity is higher when NDVI values are close to the minimum, indicating an abundance of dry vegetation when most fires occur. The fire count graph informs about the fire season timing and length, which is based on the average fire count values derived from the MODIS active fire time-series (2000–2016). Knowing when the fire season starts, and its duration, allows anomalies to be identified in any specific year. This type of analysis can be reproduced for each protected area of the WDPA and results can be discussed in the national and regional context. Comparison of fire activity with other environmental and climate indicators gives a better understanding of fire–vegetation dynamics and enables managers to identify

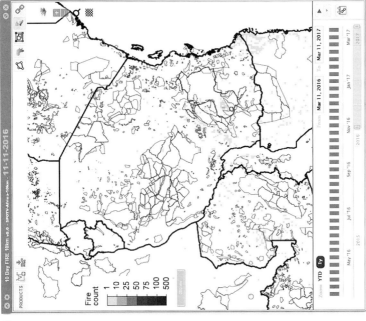

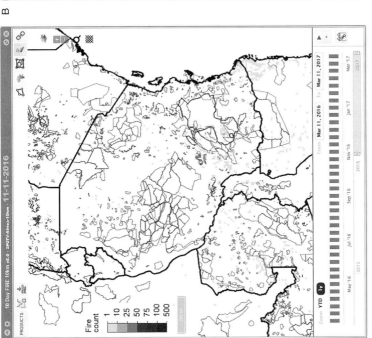

FIGURE 5.3   The information displayed is taken from the eStation user interface. (A) The map shows the total fire count in Tanzania. Fire count is derived from the MODIS active fire product over a 10-km grid, values refer to the period November 2016 to March 2017. The WDPA boundaries are displayed in dark grey. (B) The bar chart shows the MODIS active fire and NDVI values (from SPOT Vegetation imagery) in the Selous reserve during the latest fire season (May 2016–February 2017). Values refer to 10-day periods. Bars show the fire count (in light grey, the current season; in dark grey, the average value from the 2000–2016 period), lines represent the NDVI term (in dark grey, the mean in the current season; in light grey, the mean values for the period 1999–2014).

anomalies and pressures on ecosystems. In particular, rainfall and vegetation patterns in the months preceding the start of fire activity are key drivers of fire occurrence.

The example in Figure 5.4 shows 10-day periods of rainfall, the NDVI, and the fire-count anomaly in the African countries, to demonstrate how the eStation can be used for regional and continental assessments.

These examples of analysis are the result of long-term efforts, not only in the development of the eStation platform, but also in the distribution of the services to users.

Protected Areas in Western Africa

In 2011, a collaboration between the JRC and park ecologists in western Africa (Niger and Benin) was established to improve conservation efforts and protected-area management, as well as to advance scientific knowledge on African ecosystems, based on satellite data, *in situ* measurements, and local knowledge (e.g. traditional fire practices). The JRC supplied park ecologists with regular satellite-derived information on fire activity and vegetation dynamics, in order to assess pressures and threats in the protected areas. This collaboration highlighted the needs of the ecologists who were dealing with conservation issues and provided important insights to the design of the FMT. The start of the BIOPAMA project set up the framework for the FMT full development, which became a decision-support system for global conservation areas. The ongoing MESA project also contributed to the FMT development by providing data and computing capacity.

Here, we show how the FMT was adopted in Park W (Niger), which is part of the WAP complex. WAP stands for W–Arly–Pendjari, which are the three national parks of the complex: Park W is a transnational park, located in Benin, Burkina Faso, and Niger; Park Arly is a total faunal reserve in Burkina Faso; and Pendjari is a national park in Benin. Several other protected areas and hunting reserves lie within the complex. With an area larger than 30,000 km², this complex is the largest contiguous protected area in western Africa, and it

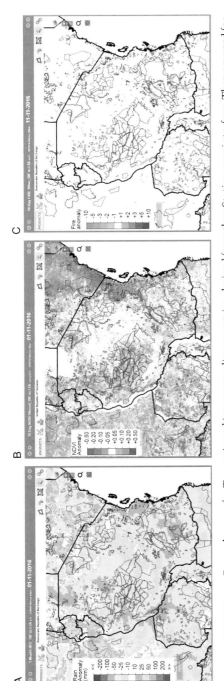

FIGURE 5.4 Example maps for Tanzania and its surrounding countries, derived from the eStation user interface. They are used for the continental assessment of wildfire, rainfall, and vegetation anomalies. The maps refer to the period 1–11 November 2016, with the WDPA boundaries displayed in dark grey. (A) The rain anomaly as derived from FWES RFE (source: NOAA's Climate Prediction Center; Xie et al., 1997). (B) The NDVI anomaly calculated as the current NDVI/average NDVI (1999–2014). (C) The fire anomaly (on a 10-km grid) based on the MODIS active-fire product (1-km).

(A black and white version of this figure will appear in some formats. For the colour version, please refer to the plate section.)

represents a key conservation area as it hosts more than half of west Africa's elephant population, and has a rich biodiversity and intact habitats (UNDP 2004).

Today, the parks have well-established fire-management plans to achieve conservation targets. Fire is used to shape the landscape, to facilitate tourist access and mobility in the park, and to maintain the ecological balance between the tree, shrub, and grass communities. Perennial grasses, with an average height of 2-m, are very common and particularly important in the park as they are a key component of ecosystem structure and fire dynamics.

Park ecologists estimate that without fire management in wes-tern-African parks, around 50 per cent of the park area could be destroyed during the dry season by the fires caused by poachers (unpublished work). Since 2005, the park rangers of this region burn around 20 per cent of the total area every year to provide fresh grass for herbivores, reduce bush encroachment, and facilitate park accessibility for tourists. Management uses three types of fires: early-, mid-, and late-season fires. The timing of the prescribed burning is crucial to achieve management objectives. More information on the burning strategies is provided in Box 5.2.

An example of management fires applied in Park W (Niger) is shown in Figure 5.5. This is a typical early burning that is applied to stimulate grass regrowth during the early dry season (Figure 5.6); fresh grass is important for the herbivores of the park and the overall animal community.

Vegetation and wildlife benefit from the planned burning prac-tices, and the overall biodiversity conservation is improved. Herbivores tend to occupy areas where vegetation is regenerating after burning, in order to graze the fresh grass growth. Antelopes and buffaloes, in particular western hartebeest (*Alcelaphus buselaphus*) and Senegal hartebeest (*Damaliscus lunatus korrigum*), use these areas. Other animals are drawn to these areas too. For example, pre-datory birds are often seen close to fire fronts, where they can hunt the animals escaping the fire.

BOX 5.2   **Management fires in West Africa**

EARLY SEASON FIRES

Early, dry-season fires tend to mainly affect grasses, with limited or no effect on trees. These fires are applied to protect park resources against large fires by creating gaps in the vegetation. They are also used to stimulate grass regrowth and provide fresh forage to herbivores during the dry season; they remove unpalatable plant species and pests, and help to control invasive plants as well. They are lit just after the end of the rainy season, when vegetation is still wet, so that grass will grow soon after the fire (i.e. around 2 weeks after the burning). Because woody vegetation is usually not damaged, in the long term these fires can create a closed savanna. The typical period to light these fires is September–October.

MID-SEASON FIRES

Fires later in the season reach higher intensity due to the drier and hotter weather conditions and the lower water content of vegetation. These fires are more destructive for young trees and shrubs (Govender *et al.* 2006).

In western Africa, the mid-season fires occur in December and January. These are the most important for the management of the park, and they are ignited during the coolest hours (usually overnight) to prevent fire reaching other parts of the park. They are important to keep the balance between the woody and herbaceous vegetation. Perennial grasses, such as *Andropogon gayanus* or *Hyparrhenia* spp., regrow quickly after the passage of fire. They provide fresh grass for herbivores during the long dry season, with effects on the wildlife movement in the park, and thus tourism routes.

BOX 5.2 (**cont.**)

LATE-SEASON FIRES

Late fires take place after the first rain (May and June) and are used only in areas of closed savanna, to destroy the dry bedding that stifles young regrowth. These types of fires, repeated over several years, create open savanna. They are rarely applied in western-African protected areas; the practice is more common in southern Africa, where it is used to reach higher combustion intensities, to limit tree growth and control bush encroachment (Laris 2011).

Park W adopted the FMT to achieve specific conservation targets that are part of the park management plan. The conservation objectives are (i) to enable early warning and intervention of poaching, (ii) to control threats and pressures inside and around the protected area, (iii) to support decision-making and planning, and (iv) to verify the fire plans.

Besides the achievement of these targets, the use of the FMT also allows the reduction of park-management costs, because it is possible to perform the ecological assessment of the park with a lower number of field surveys and personnel on the ground. Field surveys are usually the main source of information about habitat status and threats, but these can be expensive, especially in large parks where many rangers, vehicles, and equipment are needed. With the information provided by the FMT, patrolling activity and intervention (e.g. law enforcement) can be planned on the basis of conservation priority zonation and resource availability (objective (i) above). The ecological indicators and maps of fire occurrence provided by the FMT allow prompt intervention, when resources are available, and can be used to identify park vulnerabilities and threats (objective (ii) above). The FMT provides information about the typical fire-season timing, which facilitates planning operation in the park (objective (iii) above). Moreover, the assessment of the fire activity at the end

FIGURE 5.5 Management fires in Park W (Niger). The fire was lit
in November 2010 to remove the dry vegetation and stimulate growth of
new grass for herbivores.
(Photo by Karim Samna).

of the fire season can be used to evaluate and improve management
plans. For example, the estimates of burned areas derived from satellite observation can be compared to the area that should be burned
each year through management fires, to evaluate if fire plans meet
their targets (objective (iv) above).

An example of the FMT portal is shown in Figure 5.7A.
The view shows the fire activity in Park W (Niger) during the
2015–2016 fire season. The portal provides the land-cover type in
the selected site (pie chart on the top left), and the distribution of
the area burned in the main vegetation classes (pie chart on the top
right). The left panel displays statistics and indicators of the active-
fire data, whereas the right window shows information about the
extent of area burned. The tables report the fire density and the
burned area extent inside Park W, and in its 25-km buffer.
The graphs below the tables report the cumulative values of fire

FIGURE 5.6 Plant regrowth a few days after the fire. (Photo by Karim Samna).

count (left side) and burned area (right side) during the whole fire season. The bottom graphs compare the current fire-season parameters of active fires and burned areas with the average values (2003–2012). The maps in the central window provide three downloadable products in shapefile and geotiff formats: fire occurrence (from MODIS active fire), fire density (from MODIS active), and burned area (from MODIS burned area). Beside the indicators and maps provided for each area of interest, the FMT includes an interactive map viewer (Figure 5.7B), where users can customise the data to display (fire, vegetation type, protected areas) and download them. Data download is possible for any selected area of interest.

A 10-year conservation programme has been launched for Park W (Niger), over the period 2017–2026. This long-term conservation plan is the result of an agreement between several countries and organisations: Niger, Benin, Burkina Faso, European Union, United Nations

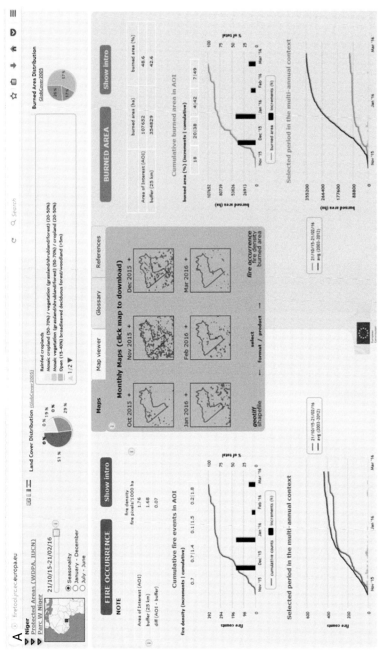

FIGURE 5.7 The FMT viewer. (A) The statistics and indicators of fire occurrence and burned areas. (B) The interactive map viewer.

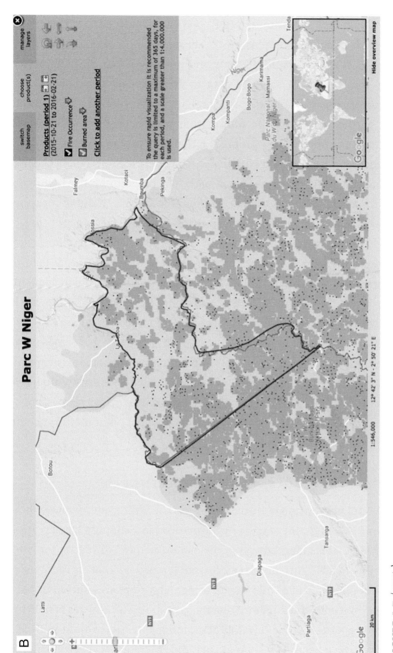

FIGURE 5.7 (cont.)

155

Development Programme (UNDP), and GIZ (Deutsche Gesellschaft für Internationale Zusammenarbeit). The programme defines the objectives of fire management inside the protected area. Importantly, it calls for the use of satellite data to complement ground-based fire monitoring, patrolling, and post-fire surveys (i.e. management objective verification). Satellite information will be used to produce estimates of the burned areas, to map active fires, and to assess the effects of burning. The satellite estimates will be a verification tool for the fire-management plans adopted in the park. The programme also highlights the importance of and the need for remote sensing training for those who coordinate ecological monitoring in the park.

Thus, the eStation and FMT can support this programme and increase the capacity of the park to use satellite and spatial data to inform ground-patrol surveys and improve fire management. Moreover, the adoption of these systems by the programme could reduce the park management costs (e.g. field ecological surveys and patrolling), as habitat monitoring can, in large part, be performed using satellite products and indicators. The mapping and analysis capabilities of both tools could be a valuable means to identify vulnerable areas and improve intervention on the ground, as priority areas can be identified and monitored more easily and conservation efforts could become more effective.

5.6   LESSONS LEARNED AND CHALLENGES

The implementation of the eStation and the FMT systems shows how remote sensing and satellite information can support conservation over large areas, and how systems can be adapted to respond to site-specific needs. An important lesson learned is that system users should play a crucial role in the design and implementation of platforms. Without the direct involvement of the communities benefiting from the systems, the development and application would not have reached the current level of maturity and efficiency. Capacity building efforts have a very important role in making the user community benefit fully from tools, and even more capacity should be developed to maximise benefits.

The central challenges encountered in using the eStation products and the FMT are related to user capacity and technology constraints.

User Capacity

Basic expertise in spatial data use and analysis are needed for the appropriate use of the tools and the information they provide. Both of the systems discussed in this chapter deliver derived products and indicators rather than raw satellite data, which make them easier to use. However, these products still require users to understand the ecological meaning of the products, and their limitations. For example, the spatial and temporal resolution of the datasets has to be considered in relation to the geographic scale of analysis. Indicators based on very low spatial resolution might not be appropriate or could strongly limit the analysis of processes at the local level (e.g. protected area). Additionally, the interpretation of the results requires a full understanding of the product feature used for the assessment. Another element affecting the success of these products is the user expertise in spatial data interpretation and analysis. Some users will use data to produce reports and bulletins, while others will use these data for more advanced analysis, to generate maps and statistics.

A prior knowledge of GIS and spatial datasets will greatly increase the benefit derived from the use of these tools. The MESA project developed a long-term training programme for eStation products and tools. This was extended to cover data processing, algorithm development, and software engineering for staff responsible for the eStation server. Other training was dedicated to environmental indicators that were used in the reporting and analysis of wildfire, agriculture, and climate data. The MESA training programme will continue in the next years, and more information can be found at http://training4mesa.org. Most activities are organised in the MESA training centres in Africa and Europe (i.e. EUMETSAT and JRC premises), but some training is delivered through the EUMETCast channel as distance education courses. These are usually given in English or French, depending on the participant's preference. Training spans multiple types of users (e.g. decision-makers, researchers) and therefore the courses are diverse. Some include the analysis of remote sensing data and the use of environmental information in policy formulation and decision-making. Other courses focus on specific thematic areas of application (e.g. climate, vegetation, wildfire etc.) to increase the capacity in satellite data analysis for environmental and ecological assessment. Finally, some courses are dedicated to trainers, to create more expertise in the regions and to reinforce local capacity development activities.

The FMT entry page supplies a tutorial for users that can be accessed at http://acpobservatory.jrc.ec.europa.eu/content/fire-monitoring. Dedicated training was organised for ecologists in the WAP complex and other users in conjunction with MESA training sessions about fire ecology and monitoring. Ideally, training on the use of the FMT interface and products should be increased among conservation practitioners, in order to facilitate its use for habitat monitoring, data analysis, and support to decision-making, more widely. This would enable more users to utilise the datasets about fire activity, to produce customised maps, or for further analysis when downloaded to the local GIS environment. Increasing the expertise and capacity of the park staff in interpreting spatial data and the derived indicators will make for more effective management of fire within and around their protected areas. Importantly, training programmes offer the participating staff networking opportunities that could facilitate information exchange and collaboration. This can create valuable networks of experts who can provide mutual support and assistance to resolve management and ecological issues. Lessons learned in a protected area may be shared and replicated in others. The creation of such networks among conservation practitioners will also widen the local conservation efforts, for example by establishing regional ecological corridors between protected areas, to increase habitat connectivity. This will eventually enhance the impact of local conservation policies, for instance by facilitating the control of illegal activities, such as poaching, thanks to the faster communication between patrolling teams.

Technology Constraints

The eStation data delivery relies on the EUMETCast broadcasting system and the information comes only in part over the internet. This allows users with limited or unstable internet to receive the service regularly. Most eStations are installed in the ministries or regional centres for climate, forestry, and agriculture. Therefore, they do not directly reach conservation professionals working in protected areas or remote places. The use of the eStation services in these remote locations could be improved through the creation of connections between the institutions where the eStation is available and the users who require the products, whether they are universities or nongovernmental organisations dealing with conservation research or implementation activity. On a more technical level, the eStation operating system requires specialised maintenance and updating. This is provided by the information technology (IT) staff working on-site and, if necessary, the JRC staff provide assistance remotely or through a direct visit to the region. This represents a challenge, but the experience gained in the past years has made system maintenance easier.

One of the major limiting factors affecting the use of the FMT in Africa is internet access; reliable broadband is not always available, especially in protected areas. Again, this could be resolved by establishing close collaborations between the institutions with reliable access to the FMT and protected areas, or other end-users that require the service but cannot access it due to limited broadband or lack of IT infrastructure. Communication could be done through mobile phones to create a near-real-time alert system for early warnings. A similar communication network could be established between those organisations receiving the eStation services and other institutions dealing with conservation and sustainable resource management, to supply the service where internet is not available. Ideally, the eStation system and the FMT should be used together, not only because they provide complementary information for habitat monitoring, but also to overcome some of the technical limitations, such as the lack of broadband or IT infrastructure.

REFERENCES

Archer, A. J. (2000). *Pinus serotina*. Fire Effects Information System. US Department of Agriculture. See www.feis-crs.org/feis/.

Arno, S. F. (1996). The seminal importance of fire in ecosystem management. In *The Use of Fire in Forest Restoration: USDA Forest Service General Technical Report INT-GTR-341*. Ogden, UT: Intermountain Research Station.

Barbosa, P., Stroppiana, D., Grégoire, J.-M., and Pereira, J. M. C. (1999). An assessment of vegetation fire in Africa (1981–1991): burned areas, burned biomass, and atmospheric emissions. *Global Biogeochemal Cycles*, **13**, 933–950.

Bond, W. J., Woodward, F. I., and Midgley, G. F. (2005). The global distribution of ecosystems in a world without fire. *New Phytologist*, **165**, 525–538.

Boschetti, L., Roy, D. P., Justice, C. O., and Humber, M. L. (2015). MODIS–Landsat fusion for large area 30 m burned area mapping. *Remote Sensing of Environment*, **161**, 27–42.

Bowman D. M. J. S., O'Brien, J. A., and Goldammer J. G. (2013). Pyrogeography and the global quest for sustainable fire management. *Annual Review of Environment and Resources*, **38**, 57–80.

Clerici, M., Combal, B., Pekel, J. F., *et al.* (2013). The eStation, an Earth observation processing service in support to ecological monitoring. *Ecological Informatics*, **18**, 162–170.

Davies, D. K., Ilavajhala, S., Wong, M. M. and Justice, C. O. (2009). Fire Information for Resource Management System: archiving and distributing MODIS active fire data. *IEEE Transactions on Geoscience and Remote Sensing*, **47**, 72–79.

Dwyer, E., Pereira, J. M. C., Grégoire, J-M., and Dacamara, C. C. (1999). Characterization of the spatio-temporal patterns of global fire activity using satellite imagery for the period April 1992 to March 1993. *Journal of Biogeography*, **27**, 57–69.

Dwyer, E., Pinnock, S., Grégoire, J. M., and Pereira, J. M. C. (2000). Global spatial and temporal distribution of vegetation fire as determined from satellite observations. *International Journal of Remote Sensing*, **21**, 1289–1302.

Eva, H. D., Malingreau, J. P., Gregoire, J. M., and Belward, A. S. (1998). The advance of burnt areas in Central Africa as detected by ERS-1 ATSR-1. *International Journal of Remote Sensing*, **19**, 1635–1637.

Giglio, L., Kendall, J. D., and Justice, C. O. (1999). Evaluation of global fire detection algorithms using simulated AVHRR infrared data. *International Journal of Remote Sensing*, **20**, 1947–1985.

Giglio, L., Descloitres, J., Justice, C. O., and Kaufman, Y. (2003a). An enhanced contextual fire detection algorithm for MODIS. *Remote Sensing of Environment*, **87**, 273–282.

Giglio L., Kendall J., and Mack R. (2003b). A multi-year active fire dataset for the tropics derived from the TRMM VIRS. *International Journal of Remote Sensing*, **24**, 4505–4525.

Govender, N., Trollope, W., and Van Wilgen, B. (2006). The effect of fire season, fire frequency, rainfall and management on fire intensity in savanna vegetation in South Africa. *Journal of Applied Ecology*, **43**, 748–58.

Gerard, F., Plummer, S., Wadsworth, R., et al. (2003). Forest fire scar detection in the boreal forest with multitemporal SPOT-VEGETATION data. *IEEE Transactions on Geoscience and Remote Sensing*, **41**, 2575–2585.

Grégoire, J-M., Tansey, K., and Silva, J. M. N. (2003). The GBA 2000 initiative: developing a global burned area database from SPOT Vegetation imagery. *International Journal of Remote Sensing*, **24**, 1369–1376.

Hardesty, J., Myers, R. L., and Fulks, W. (2005). Fire, ecosystems, and people: a preliminary assessment of fire as a global conservation issue. *The George Wright Forum*. **22**, 78–87.

Komarek, E. V. (1983). Fire as an anthropogenic factor in vegetation ecology. In Holzner, W., Werger, M. J. A., and Ikusima, I., eds., *Man's Impact on Vegetation*. Boston, MA: Dr W. Junk Publishers: pp. 77–82.

Kaufman, Y., Remer, L. Ottmar, R., et al. (1996). Relationship between remotely sensed fire intensity and rate of emission of smoke: SCAR-C experiment. In Levine, J., ed., *Global Biomass Burning*, Cambridge, MA: MIT Press, pp. 685–696.

Kasischke, E. S., Bergen, K., Fennimore, R., et al. (1999). Satellite imagery gives clear picture of Russia's boreal forest fires. *Eos, Transactions American Geophysical Union*, **80**, 141–147.

Laris, P. (2011). Humanizing savanna biogeography: linking human practices with ecological patterns in a frequently burned savanna of southern Mali. *Annals of the Association of American Geographers*, **101**, 1067–1088.

Li, F., Bond-Lamberty, B., and Levis S. (2014). Quantifying the role of fire in the Earth system – Part 2: impact on the net carbon balance of global terrestrial ecosystems for the 20th century. *Biogeosciences*, **11**, 1345–1360.

Myers, R. L. (2006). *TNC Living with Fire: Sustaining Ecosystems & Livelihoods Through Integrated Fire Management*. Tallahassee, FL: The Nature Conservancy.

Page, S. E. and Hooijer, A. (2016). In the line of fire: the peatlands of Southeast Asia. *Philosophical Transactions of the Royal Society B*, **371**, doi: 10.1098/rstb.2015.0176.

Pereira, J., Pereira, B. S., Barbosa, P., *et al.* (1999). Satellite monitoring of fire in the EXPRESSO study area during the 1996 dry season experiment: Active fires, burnt area, and atmospheric emissions. *Journal of Geophysical Research: Atmospheres*, **104**, 30701.

Roy, D. P., Jin, Y., Lewis, P. E., and Justice, C. O. (2005). Prototyping a global algorithm for systematic fire-affected area mapping using MODIS time series data. *Remote Sensing of Environment*, **97**, 137–162.

Roy, D. P., Boschetti, L., Justice, C.O., and Ju, J. (2008). The Collection 5 MODIS Burned Area Product – global evaluation by comparison with the MODIS Active Fire Product. *Remote Sensing of Environment*, **112**, 3690–3707.

Roy, D. P., Boschetti, L., and Smith, A. (2013). Satellite remote sensing of fires. In Belcher, C. M., ed., *Fire Phenomena and the Earth System: An Interdisciplinary Guide to Fire Science*. Hoboken, NJ: Wiley-Blackwell, pp. 77–93.

Scholes, R. J. and Archer, S. R. (1997). Tree-grass interactions in savannas. *Annual Review of Ecology and Systematics*, **28**, 517–544.

Schroeder, W., Oliva, P., Giglio, L., and Csiszar, I. A. (2014). The new VIIRS 375 m active fire detection data product: algorithm description and initial assessment. *Remote Sensing of Environment*, **143**, 85–96.

Schultz, M. G. (2002). On the use of ATSR fire count data to estimate the seasonal and interannual variability of vegetation fire emissions. *Atmospheric Chemistry and Physics*, **2**, 387–395.

Shlisky, A., Waugh, J., Gonzalez, P. *et al.* (2007). Fire, ecosystems and people: threats and strategies for global biodiversity conservation. *Global Fire Initiative Technical Report 2007–2*. Arlington, VA: The Nature Conservancy.

Sudhakar Reddy, C., Padma Alekhya, V. V. L., Saranya, K. R. L., *et al.* (2017). Monitoring of fire incidences in vegetation types and protected areas of India: implications on carbon emissions. *Journal of Earth System Science*, **126**, 11.

Tansey K., Grégoire J-M., Stroppiana D., *et al.* (2004). Vegetation burning in the year 2000: global burned area estimates from SPOT Vegetation data. *Journal of Geophysical Research*, **109**.

Tansey, K., Grégoire, J.-M., Defourny, P., *et al.* (2008). A new, global, multi-annual (2000–2007) burnt area product at 1 km resolution. *Geophysical Research Letters*, **35**, L01401.

Turner, M. G., Romme, W. H., and Tinker, D. B. (2003). Surprises and lessons from the 1988 Yellowstone fires. *Frontiers in Ecology and the Environment.* **1**, 351–358.

UNDP (United Nations Development Programme) (2004). Enhancing the effectiveness and catalyzing the sustainability of the W-Arly-Pendjari (WAP) protected area system. UNDP Project Document PIMS 1617.

UNEP (United Nations Environment Programme) and IUCN (2010). *The World Database on Protected Areas (WDPA): Annual Release 2010.* Cambridge: United Nations Environment Programme World Conservation Monitoring Centre.

Verhegghen, A., Eva, H. Ceccherini, G., *et al.* (2016), The potential of Sentinel satellites for burnt area mapping and monitoring in the Congo Basin forests. *Remote Sensing,* **8**, 986.

Xie, P. and Arkin, P. A. (1997). A 17-year monthly analysis based on gauge observations, satellite estimates, and numerical model outputs. *Bulletin of the American Meteorological Society,* **78**, 2539–2558.

Zhang, Y. H., Wooster, M. J., Tutubalina, O., and Perry, G. L. W. (2003). Monthly burned area and forest fire carbon emission estimates for the Russian Federation from SPOT VGT. *Remote Sensing of Environment,* **87**, 1–15.

6 Ecosystem Functioning Observations for Assessing Conservation in the Doñana National Park, Spain

Paula Escribano and Néstor Fernández

6.1 INTRODUCTION

Protected areas are a cornerstone of biodiversity conservation. They are critically important, not only for preserving endangered species and their habitats, but also for delivering essential ecosystem services (Watson *et al.* 2014). Protected areas currently cover around 12 per cent of the Earth's surface, and countries are on track to increasing these areas to up to 17 per cent of the terrestrial surface, as agreed by the Convention on Biological Diversity (CDB) under the Strategic Plan for Biodiversity 2011–2020 (Aichi target 11). However, several examples have demonstrated that protection is not necessarily sufficient to guarantee the ecosystem integrity or to preserve biodiversity (Costelloe *et al.* 2016). To better maintain, and even improve, the conservation performance of these areas, managers, decision-makers, and stakeholders require precise information on the natural dynamics of ecosystems, trends, and disturbances. Therefore, there is an increasing need for affordable and easily accessible monitoring tools, capable of providing relevant ecological information for decision-making (Worboys *et al.* 2015, López-Rodríguez *et al.* 2017).

The application of satellite remote sensing in conservation science has dramatically increased our capacity to assess management effectiveness in protected areas. Remote sensing is now extensively used for mapping and monitoring protected habitats since it provides affordable, spatially continuous, and temporally comprehensive data. Very often, remote sensing is used to monitor structural, 'tangible' components of protected areas, for example in conducting vegetation mapping, monitoring vegetation succession, identifying disturbances such as urbanisation, logging, fires, etc., or mapping the distribution of invasive species (e.g. Nagendra *et al.* 2013). Complementary monitoring programmes, focusing on ecosystem functioning such as the dynamics of primary productivity, carbon assimilation, and water fluxes are also required for many important management targets, for example assessing the vulnerability of ecosystems to climate change (Huete 2016) or supporting the evaluation of ecosystem services that protected areas provide (Cabello *et al.* 2012).

Climate change imposes new management challenges on protected areas. In these areas, the boundaries are fixed whilst climate evolves in time. Therefore, protected areas need to be managed to increase resilience in the presence of new climatic conditions. In order to maintain their structure, composition, and function, the CBD has encouraged the implementation and planning of ecosystem-based approaches, to reduce vulnerability and to increase the resilience of ecosystems to climate change (Convention on Biological Diversity 2010). The impacts of climate change may vary globally. Mediterranean regions have been highlighted as being particularly vulnerable to climate change (Giorgi 2006, Baettig *et al.* 2007). Therefore, local management actions to reduce anthropogenic impacts on ecosystems could be particularly important to ameliorate the effects of climate change. They might also benefit ecosystem sustainability and ensure that the ecosystem can continue to supply benefits and goods for people.

Even though managing ecosystems for increasing resilience is a major concern, as Munby and Anthony (2015) noted there are a lack of operational methods that can be utilised to actually improve resilience. Furthermore, the definition of resilience is often inconsistent across the scientific literature, and it is often difficult to translate concepts and methods for actually measuring resilience from one area to another. This lack of clarity makes it even more difficult to provide managers with easily accessible information on ecosystem dynamics and changes (Munby and Anthony 2015). Therefore, now, more than ever, it is necessary to make an effort in translating complex data and analyses into outputs that are easily exchangeable between different stakeholders, to bridge the gap between science and management.

In this chapter, we describe our experience in developing an observatory of ecosystem function, based on satellite remote sensing data in Doñana National Park. We wanted to provide practitioners with accessible information on ecosystem changes, trends, and responses to disturbances, specifically to extreme drought events and management actions in terms of carbon and water balance. This observatory was intended as a first step in the joining of science and management, and to understand the gaps in knowledge and/or implementation for evidence-based management practices. Specifically, the aim of the observatory was to offer a platform to facilitate the communication between scientists and policy-makers, translating complex scientific data and analyses into simple graphics and visualisations. This platform provides information that can help in different management aspects by (i) introducing ecosystem functioning into monitoring programmes, (ii) introducing future scenarios of climate change into management, and (iii) monitoring the consequences of the management actions in terms of carbon and water balance. The methodology and tools developed and the knowledge gained during the implementation of this observatory is an example that could be transferred to other protected areas.

6.2   THE DOÑANA NATIONAL PARK: ENVIRONMENTAL CONTEXT AND PRESSURES

The Doñana National Park (Figure 6.1) is a United Nations Educational, Scientific and Cultural Organization (UNESCO) World Heritage site and is one of the most emblematic protected areas in Europe. The Park was established in 1969 in response to the extensive conversion of unique marshland and shrubland ecosystems into crops and forest plantations. The protected area was subsequently expanded to the current extent of 54,000-ha, plus an additional buffer area of

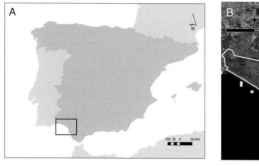
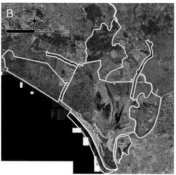

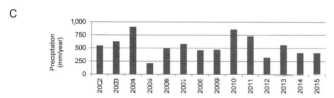

FIGURE 6.1 Study area: (A) The Iberian Peninsula, (B) Doñana National Park, and (C) the mean annual precipitation.

68,000-ha. Together, the protected areas have preserved unique and highly diverse landscapes, including natural and cropped wetlands, temporal lagoons, Mediterranean shrublands and forests, and coastal dune systems. With this has come the preservation of high species richness, including highly charismatic and endangered terrestrial vertebrates such as the Iberian lynx (*Lynx pardinus*) and the Spanish imperial eagle (*Aquila adalberti*). The region is of special conservation importance in both the Mediterranean and European contexts. The Doñana wetlands also constitute one of the most important wintering areas for waterbirds in Western Europe (Rendón 2008). The climate in Doñana is Mediterranean subhumid, with mild and wet winters and hot and dry summers, and with an average precipitation of 550-mm/year (Figure 6.1C). The summer water deficit imposes the major climatic control on vegetation activity due to the co-occurrence of water shortage and daily maximum temperatures (usually > 30 °C) between June and September.

Conservation Threats

There is an increasing concern about the risk of collapse of some of the most valuable areas within Doñana National Park, as the intensification of agricultural practices and the exploitation of water resources in surrounding regions continue to grow (Green *et al.* 2017). Shifts in the hydrologic regime of temporal lagoons have already been observed, including reductions in the annual flooding period that have been associated with groundwater depletion (Gómez-Rodríguez *et al.* 2010). These shifts have already caused declines in the cover of hydrophitic vegetation and reductions in plant species richness in temporary lagoons (Serrano and Zunzunegui 2008). The current flooding regime of the natural marsh is mostly determined by local precipitation, as the main tributary rivers were channelled during the twentieth century for the conversion of large sectors of marshland into cropland. The dependence of the remaining natural marsh on local precipitation makes the flooding regime and marsh vegetation particularly sensitive to climate change. Furthermore, the conservation status of Doñana is threatened by groundwater extraction for agriculture and mass tourism in the immediate surroundings of the protected area. These threats have contributed to the ranking of this World Heritage area as under 'very high threat' according to the IUCN (www .worldheritageoutlook.iucn.org/).

In addition, extensive livestock grazing is allowed within the protected area. Livestock are considered to be a traditional, sustainable practice in management plans, but are regulated through the adjustment of carrying capacities. Grazing and trampling by domestic herbivores can have a strong impact on vegetation and on the habitat of other species (Soriguer *et al.* 2001). The incidence of these impacts can be difficult and expensive to monitor over large spatial extents using traditional field methods.

Drought is the main factor controlling vegetation growth in Mediterranean areas (Gouveia *et al.* 2017). These areas typically suffer long summers in which lack of water, and high irradiance and temperature levels, are coupled. The typical hydrological cycle includes a sequence of wet–dry years. However, new climate-change scenarios predict an increase in the length and intensity of dry periods (IPCC 2012), making Mediterranean ecosystems increasingly vulnerable to climate change (Gouveia *et al.* 2017). Despite the overwhelming evidence of the devastating effects of increased drought on terrestrial productivity, not only in Mediterranean regions but also across Europe (Ciais *et al.* 2003), and worldwide (Huang *et al.* 2016), there is still a lack of understanding, regionally and locally, about how and to what extent drought affects different ecosystems. There is an urgent need to understand the impact of an increase in drought at different spatial levels and to develop tools useful for sustainable management of protected areas under the threat of climate change.

6.3 MONITORING ECOSYSTEM FUNCTIONING AT THE ECOSYSTEM OBSERVATORY OF DOÑANA

In the Mediterranean region, vegetation has evolved to cope with long summers of high temperatures and low precipitation rates. These conditions result in rapid primary productivity decays and increases in evapotranspiration rates. Variations in this pattern are also observed in association with differences in landform and groundwater availability (e.g. between shrublands and forests in sandy soils, and wetlands in depressions, Fernández *et al.* 2010). The effects of the water deficit on the dynamics of primary productivity have important consequences for primary consumers, determining the quality of their habitats and, ultimately, their population dynamics (Fernández *et al.* 2016). Therefore, it is important to understand (i) the dynamics of primary production or the gains and losses of carbon, and (ii) the dynamics of vegetation water deficit. These are the two key descriptors of ecosystem functioning upon which our analysis of remote sensing data for the observatory are based.

Primary productivity is a key component of carbon balance on the Earth's surface as it determines the amount of carbon that is assimilated by vegetation through photosynthesis. We used the

Enhanced Vegetation Index (EVI; Huete *et al.* 2002) as a proxy of terrestrial productivity for its direct relation with the green-leaf-area index, green biomass, and the percentage of green vegetation (Baret and Guyot 1991).

Evapotranspiration measures the amount of water that is returned to the atmosphere by soil transpiration and by transpiration of plant leaves. Almost 70 per cent of the water that an area receives from precipitation is lost (returned to the atmosphere) by evapotranspiration, and therefore this process is a key variable to understand the water cycle. We developed a time-series of water deficit for the entire protected area based on the Temperature–Vegetation Dryness Index (TVDI; Sandholt *et al.* 2002). TDVI was used as proxy of evapotranspiration as proposed by Garcia *et al.* (2014). The calculation of this index requires solely information derived from remotely sensed data, specifically the EVI and land surface temperature.

Deciding on the Most Appropriate Sensor

Given the breadth of satellite sensors and data products now available (Chapter 2), determining which sensor is the most appropriate in relation to the monitoring needs is an important decision. Ideally, analyses of trends and changes in ecosystem functioning will use datasets with a long historical record (more than two decades), taken at regular time intervals and with high temporal resolution (an image per day would be optimal). The Moderate Resolution Imaging Spectroradiometer (MODIS) sensor is at present the most suitable sensor for devising affordable, long-term monitoring systems for the following reasons:

(i)   Spectral range. MODIS covers the visible, the near-infrared, and the thermal range, allowing for calculations of EVI and TVDI as key indicators of carbon and water fluxes.

(ii) Balance between temporal and spatial resolution. With a revisit frequency of 1 day for the entire Earth (for each of the two sensors), MODIS is especially suited to calculate intra-annual variations in vegetation productivity and vegetation dryness. For the observatory, we selected the composite products of the MODIS sensor retrieved from daily images after a process of quality checking, which provides consistent temporal and spatial data, and is less affected by the presence of clouds, aerosols, and other noise effects than the pre-processed product. In addition, MODIS offers 250-m-resolution data for the EVI and 500-m-resolution data for the land surface temperature. Although higher spatial resolution would be desirable for monitoring some specific vegetation patches, there is no sensor that combines both high spatial resolution and a similar revisit period.

(iii) Historical record. To discern inter-annual natural fluctuations from anomalies such as changes associated with perturbations, we needed data with a long historical record. MODIS observations date back to 2000. Other sensors with longer historical record such as the Advanced Very High Resolution Radiometer (AVHRR) and the Landsat series of sensors were unsuitable for our purposes, due to their low spatial and temporal resolution, respectively.

(iv) Minimal cost. Long-term monitoring programmes often fail due to the lack of appropriate funding mechanisms (Lindenmayer and Likens 2009). Once the system has been implemented, freely available satellite data can help to make monitoring less dependent on future funding.

Components of the Ecosystem Observatory of Doñana

The observatory was designed as a set of tools for downloading, pre-processing, analysing, and developing synthetic graphics and visualisations. Most of the routines were developed in R open-source code (R Core Team 2017), to improve the exchange of knowledge and to facilitate the incorporation of new data or routines to improve the observatory in the future. The MODIS vegetation index (MODIS data product MOD13Q1) and the land surface temperature products (MODIS data product MOD12) are produced on 16- and 8-day intervals, respectively. These data can be freely downloaded from NASA (https://modis.gsfc.nasa.gov/data/dataprod/). We developed an automatic routine implemented in the R statistical package (R Core Team 2017) to download and store these data so we could be assured that they were always up-to-date. In order to assure the complete and smoothed time-series needed for monitoring purposes, we developed a pre-processing routine including noise reduction, gap filling, and smoothing, based on the adapted LOESS filter (an iteratively reweighted smoothing method (Moreno *et al.* 2014)), for the correction and interpolation of data. This routine utilised the quality flags associated with the data sources.

Vegetation productivity fluctuates following the natural hydrological cycle, meaning that dry periods/years will have lower productivity than wet periods/years. This natural fluctuation is called 'seasonality'. Subtracting this seasonality from the time-series allowed us to discern natural fluctuations from anomalies.

The seasonality was calculated as the mean EVI for a particular day of the year over the EVI time-series (Keersmaecker *et al.* 2014) (Figure 6.2A). EVI anomalies (EVI_{an}) were calculated as in Eq. 6.1 (Van Ruijven and Berendse 2010, Vogel *et al.* 2012).

$$EVI_{an} = (EVI_i - EVI_{clim})/EVI_{std}, \qquad\qquad 6.1$$

where EVI_i = the EVI on day i, EVI_{clim} = the EVI climatological mean and EVI_{std} = the standard deviation of EVI.

Negative/positive EVI anomalies occur when the current EVI is lower/higher than the climatological mean (Figure 6.2B). A negative EVI anomaly indicates a lower amount of green vegetation cover than the mean value for that specific time of the year.

We defined three parameters of short-term stability: resistance, elasticity, and relative resilience (Table 6.1). We followed the definition of resistance of Keersmaecker *et al.* (2014), in which it is defined as the magnitude of change following perturbation. We defined relative resilience as the ability of the ecosystem to recover values of the pre-disturbance event (Lloret *et al.* 2011); and elasticity as the time needed to recover values close to a reference state (Grimm and Wissel 1997).

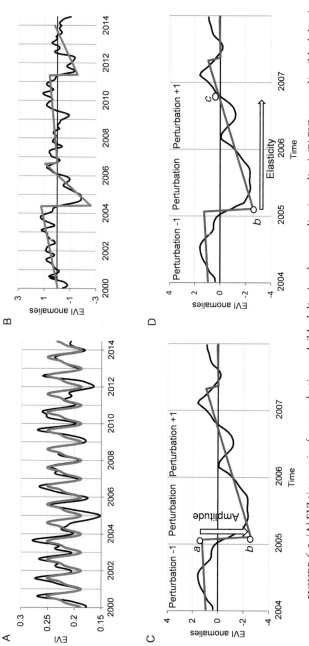

FIGURE 6.2 (A) EVI time-series for xerophytic scrub (black line) and seasonality (grey line); (B) EVI anomalies (black line) and trend component (grey line); (C) and (D) subset of EVI anomalies for the years 2004–2008, showing the data needed to calculate amplitude = $a - b$ (in the y axis, C) and elasticity = $b - c$ (in the x axis, D). See text for explanation.

Table 6.1 *State variables and stability metrics included in the observatory*

Remote sensing indicator	Indicator	Definition/interpretation
State variable	**Annual mean**	**Annual mean of anomalies** It is calculated as the annual mean for all the pixels within an ecosystem. Annual mean values of EVI_{an} close to zero indicate years in which the ecosystem has a productivity close to the climatological mean (reference state), positive/negative values indicate years with a decline/gain of productivity with respect to that reference state.
	Variance of annual mean	**Spatial heterogeneity** It is calculated as the variance of annual mean of all the pixels within an ecosystem. High EVI_{an} variance indicates a high variability of the annual mean of EVI_{an} and therefore it is used here as a measure of spatial heterogeneity.
Stability metric	**Resistance**	**The magnitude of change following perturbation** For calculating amplitude we followed the methodology of Breaks for Additive Season and Trend (BFAST, http://bfast.r-forge.r-project.org/) proposed by Verbesselt *et al.* (2010). This methodology detects the trend of the time series along the years; when this trend changes abruptly from positive (or neutral) to strongly negative it detects a negative breaking point (BP) (Figure 6.2B). The magnitude of negative BP is considered here as a proxy of 'resistance' and it is calculated as the difference between EVI_{an} values before (point a in Figure 6.2C) and after (point b Figure 6.2C) perturbation.

Table 6.1 (cont.)

Remote sensing indicator	Indicator	Definition/interpretation
		A more resistant ecosystem will show smaller differences from the seasonal mean after a disturbance (De Keersmaecker et al. 2014).
	Elasticity	**The time needed to recover to a reference state (climatological mean)** Elasticity is calculated as the time (x axis) in which the trend equals the climatological mean (point c in Figure 6.2D) after the negative BP (point b Figure 6.2D). For EVI_{an}, this indicates the moment in which productivity of the current date equals the mean productivity. Areas with higher elasticity values indicate areas that need a longer time to recover after a perturbation.
	Relative resilience	**The ability of the ecosystem to recover EVI values of pre-perturbation** It is defined as the ability of the ecosystem to recover values of pre-disturbance event. We followed the methodology proposed by Lloret et al. (2011): Relative resilience = $(EVI_{an\ (pert+1)} - EVI_{an\ (pert)})/(EVI_{an(pert-1)})$ in which $EVI_{an(pert\ +\ 1)}$ = mean annual EVI anomalies 1 year after perturbation, $EVI_{an(pert)}$ = mean annual EVI anomalies the year of perturbation, $EVI_{an(pert-1)}$ = mean annual EVI anomalies the year previous to perturbation. Areas with low values of relative resilience indicate areas in which perturbation has a long-lasting effect.

Delivering user-friendly information for Managers

A key objective of the observatory was to translate the satellite remote sensing-derived outputs into management tools that are easy to understand and can be readily interpreted for managers of Mediterranean protected areas. With this aim in mind we wanted to convert the stability metrics to graphical outputs and spatial visualisations to maximise the exchange of knowledge while maintaining simplicity. Visualisations and graphics are considered as effective tools to support science–policy interfaces (McInerny *et al.* 2014) and are easy ways of communicating science to different stakeholders. The visualisations that we use are based on:

(i) Mean–variance plots. The annual mean was used as a proxy of annual primary productivity, whilst the variance was used as a proxy of spatial heterogeneity (Table 6.1). This method assumes that a certain ecosystem will have a relatively stable behaviour in the mean–variance space. A shift in this space will occur after a perturbation. Tracking the trajectory of this shift along the years will give an idea of how much the ecosystem has changed following a perturbation. In addition, it gives information about how many years it took an ecosystem to return to the mean–variance space prior to the perturbation. This graphical analysis was first developed by Pickup and Foran (1987) and was later implemented by Washington-Allen *et al.* (2008).

(ii) Resistance–elasticity plots. Managers can visualize, at a glance, a comparison of the magnitude of change and the time needed to recover after a perturbation, with a synoptic overview of the stability of the ecosystem following a perturbation.

(iii) Spatial visualisation of stability metrics. Geospatial information is an effective way to translate 'simple values' into explicit spatial data, offering an effective platform to establish conceptual frameworks or defining the causes and consequences of these patterns.

6.4 MANAGEMENT AND SCIENTIFIC QUESTIONS

Below, we provide some examples of how the application of remote sensing tools can address pressing protected-area management and scientific questions in Doñana and in other Mediterranean protected areas.

Understanding the Effects of Severe Droughts on Carbon Dynamics at the Landscape Scale

The consequences of increasingly extreme drought events may produce spatially and temporally heterogeneous patterns (Carnicer et al. 2011). The consequences of a drought event depend on the intensity of the perturbation and the inherent vulnerability of the ecosystem (Chapin et al. 2011, IPCC 2012), meaning that different ecosystems can display differences in stability to the same perturbation event. We analysed the effects that a severe drought event had on vegetation cover for different ecosystems. This drought event occurred in 2005 and resulted in a 65 per cent reduction of mean annual precipitation (precipitation data available at http://icts.ebd.csic.es).

The mean–variance plot allowed the visualisation of the stability of different ecosystems to the same drought event. In the example illustrated in Figure 6.3, both types of shrubland suffered a decline in productivity (a decrease in mean EVI values) following the 2005 extreme drought. A year after the perturbation (2006), the xerophytic shrubland showed a continued decline of vegetation productivity. There was notable high spatial variability (high EVI variance) in recovery, as some areas (pixels) showed productivity values close to recovery, while other areas showed a more persistent drought effect. The ecosystem fully recovered its mean–variance space in productivity 2 years after perturbation. By comparison, in the hydrophytic shrublands the 2005 drought had a more persistent effect. Two years after the perturbation, the ecosystem had not fully recovered its position within the mean–variance space before perturbation.

A mean–variance plot (Pickup and Foran 1987) is an easy-to-use methodology that was utilised to convey information, in order to aid understanding of the stability of ecosystems to drought (Férnandez et al. 2016, Escribano et al. 2017). The simple rationale behind this plot facilitated the communication process as it was easy to translate

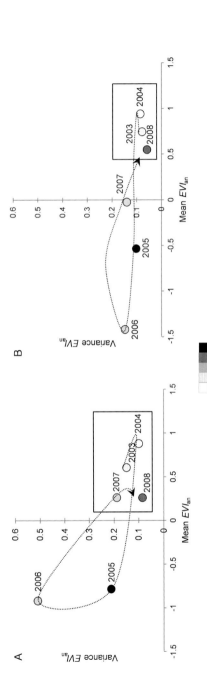

FIGURE 6.3 The mean–variance EVI space for (A) xerophytic scrub and (B) hydrophytic scrub for the years 2000–2008. This method assumes that a certain ecosystem has a stable mean–variance space (represented here as the quadrangle) and a perturbation leads to a shift in this space. Following this trajectory along the years (black dotted line) gives an overview of the changes in productivity and spatial heterogeneity following a perturbation.

shifts in the mean–variance space with field-based knowledge of vegetation decline and spatial heterogeneity.

This methodology has previously been applied to data from Landsat sensors (Washington-Allen *et al.* 2008, Cui *et al.* 2013). These authors started by selecting one image per year for their analysis, based on the best-quality image (free of clouds) around the date of maximum productivity. The longer historical record of the Landsat sensors is an advantage over MODIS; nevertheless, the revisit time of Landsat (a minimum of 16 days in a cloud-free area) can be a source of error, as the same image will not be available on the same date every year (mainly due to the presence of clouds). This is why MODIS observations were the best choice for the observatory. However, future work could combine the outputs of MODIS and higher-spatial-resolution sensors for multi scale analysis. In this way, drought effects could be detected at a high temporal resolution (MODIS) and at a high spatial resolution. This would allow for more precise identification of areas of management interest.

Monitoring Ecosystem Resilience after Extreme Drought Events

To provide baseline data to monitor resilience, we analysed the amplitude of change (the amount of productivity lost following a perturbation) and the elasticity (the number of days needed to recover plant productivity values similar to the reference state). In the example shown in Figure 6.4, we measured the changes in productivity following the severe water shortage event of 2005. We plot three different ecosystems to illustrate that the response to drought varied dramatically between the different ecosystems. Pine forests were more resistant to drought than xerophytic shrub, suffering lower decreases in biomass following the 2005 perturbation than shrublands (Figure 6.4). However, the elasticity was higher for pine forests (on average they took more than 850 days to recover vegetation cover similar to the reference state) and lower for xerophytic shrub (approximately 600 days on average). Xerophytic shrub showed a higher amplitude of change, meaning that

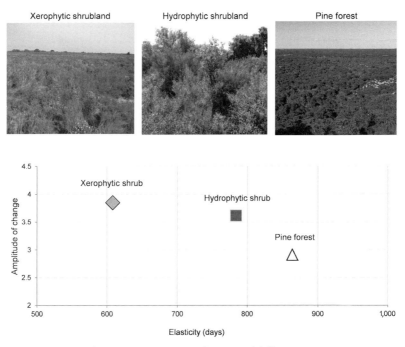

FIGURE 6.4 Resistance versus elasticity of different ecosystems.

the decline of productivity was acute after the perturbation. But this vegetation showed a high capacity to recover, a characteristic that has been observed also in other Mediterranean shrublands (Ivits *et al.* 2016).

The patterns observed from remote sensing data are supported by field studies in Doñana. These found that the 2005 drought caused extensive shrubland die-off but they also highlighted the high capacity of these shrubland communities to recover after the acute drought event (Del Cacho and Lloret 2012). De la Riva *et al.* (2016) studied the effects of drought on a set of different functional traits and found an increase of root dry matter, leaf chlorophyll, and plant height 2 years after perturbation. The support that these field-based results give to the resistance–elasticity graphics produced from remote sensing data validates the utility of these tools for remote monitoring.

Visualising these two dimensions of stability (amplitude and elasticity) has been proven to be a powerful tool for producing an overview of the ecosystem responses to drought (De Keersmaecker *et al.* 2014, Ivits *et al.* 2016). We have taken the visualisation further and shown that this output is also a powerful tool for communicating with conservation managers. It can aid their understanding of our analysis method and outputs and on the differences in stability to drought among ecosystems.

Identification of Areas with a Heightened Response to Drought

When an ecosystem is approaching a critical threshold, it can show functional divergences from the norm, and may respond more severely to perturbations (Dakos *et al.* 2008). Areas with amplified responses to drought will be more vulnerable to additional disturbances (Simoniello *et al.* 2008), in which small changes in environmental conditions can lead to major and unpredictable changes of ecosystem composition, structure, and function. It can be difficult for the ecosystem to recover once this critical transition occurs. Consequently, it is important to identify areas which are approaching 'tipping points', to prevent such critical transitions (Dakos *et al.* 2008).

In Figure 6.5 we show a section of an area neighbouring the marshland. We calculated the relative resilience (the time needed to regain vegetation-cover rates similar to those before the perturbation) of xerophytic shrublands, at a pixel scale, in order to highlight shrubland areas (pixels) with an amplified response to drought. Pixels with lower relative resilience values were areas in which the 2005 drought event had a longer-lasting effect on vegetation cover (pixels in black in Figure 6.5) than other pixels belonging to the same ecosystem.

The analyses indicate a clear spatial pattern for the relative resilience to decrease with distance to the marshland (Figure 6.5). Shifts in the hydrological regimes of temporal lagoons have already been observed in the area, including generalised reductions in the annual flooding period. This, in turn, has been associated with groundwater depletion (Gómez-Rodríguez *et al.* 2010). The higher relative resilience of the pixels around marshland can also indicate a higher

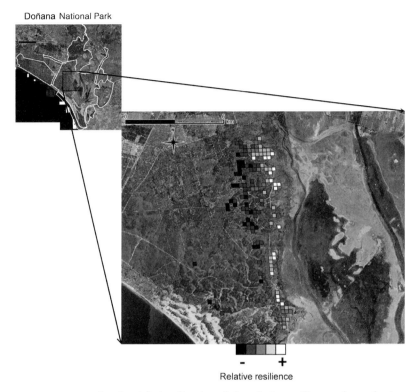

FIGURE 6.5 Spatial visualisation of the relative resilience of xerophytic shrubland.

vulnerability if a shift in the annual flooding period occurs (see Serrano and Zunzunegui 2008). Unfortunately, our remote sensing-based spatial visualisation does not allow us to investigate the local factors affecting the resilience of the shrublands, but it does highlight pixels with an anomalous behaviour to which management actions can be targeted. This management should in turn help both to understand and, ultimately, to mitigate the effects of the perturbations.

What Are the Repercussions of Management Actions in Terms of Water Balance?

Protected-area conservation requires information on the water cycle and hydrology to mitigate economic, societal, and ecological damage derived from climate change. Evapotranspiration is a key variable that links water, energy, and carbon (Monteith 1965, Fisher 2013). Doñana National Park is far from pristine. Pine forests in Doñana often consist of dense blocks of native stone pine (*Pinus pinea*), which were planted during the second half of the twentieth century. These plantations were part of the efforts to mitigate soil erosion, which had, in turn, resulted from intense clear-cutting in previous centuries. Nowadays these afforested blocks form a dense and uniform pine forest. Recently, there has been large investment made into pine forest management, to increase biodiversity of the understory of these areas with local shrub species. These efforts seek to alleviate the pressure that pine forests place on groundwater stores. A successful evaluation of the recent pine forest management activities requires a thorough assessment of changes in the water balance, linked to vegetation type.

To conduct this evaluation we monitored the effects of the clear cutting management that took place in the area between 2009 and 2010. This management aimed to encourage the regeneration of natural shrublands and to alleviate pressures on the water table. The percentage differences in evapotranspiration between managed and unmanaged areas are shown in Figure 6.6. The positive/negative differences (areas in light /dark grey respectively in Figure 6.6A) indicate days in which the evapotranspiration in the managed area exceed, or otherwise, the values observed in the unmanaged areas. Visualising these daily differences between managed and unmanaged pixels from 2000–2014 (Figure 6.6A) illustrated that, before perturbation, both areas behaved similarly in terms of evapotranspiration, but the clear-cutting event (white arrow in Figure 6.6A) provoked a strong decrease of evapotranspiration in the managed areas, with a maximum average difference of 20 per cent (May–June 2011).

The analysis and visualisation of these differences on a yearly basis (Figure 6.6B) illustrates to managers that differences in evapotranspiration are highest the year after the perturbation (in 2011). These differences in evapotranspiration tended to decrease gradually in subsequent years until evapotranspiration rates became similar

184

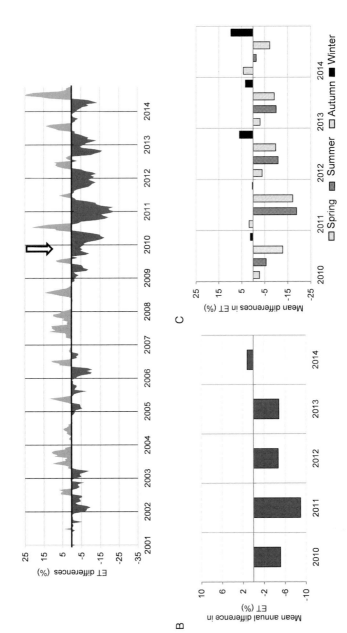

FIGURE 6.6 (A) Daily differences in evapotranspiration (ET) between managed and unmanaged areas for 2001–2014. The white arrow indicates the event of clear-cutting. (B) Mean annual differences and (C) mean seasonal differences for 2010–2014.

(2% higher for managed areas), between managed and unmanaged areas, 4 years after the clear-cutting event. Through the analysis of these data on a seasonal basis (Figure 6.6C), it became evident that clear-cutting resulted in a strong decrease in evapotranspiration in summer. This is the season in which the water demand is high in the area. These simple graphics were useful tools to evaluate the effects of management actions in water balance and they highlighted the need for regular data collection.

6.5 HISTORY AND LESSONS LEARNED

The Ecosystem Observatory of Doñana has made significant progress delivering relevant information ecosystem function for the management of Doñana National Park. Following a period of 'passive protection' after the creation of Doñana National Park, ecological restoration activities intensified in the late 1990s. The aim of these activities was to restore areas within the park that were highly degraded before the establishment and the expansion of the protected area. The activities have resulted in the restoration of the water regime in drained sections of the marsh, the removal of alien eucalyptus plantations, and the facilitation of natural vegetation succession in areas where native pine trees had been planted at unnaturally high densities. Monitoring efforts, however, have traditionally focused on structural components of biodiversity, such as animal population, vegetation structure, and the hydrological regime. The need to complement protected-area monitoring with the metrics of ecosystem functioning was first raised by scientists working in the park in the first decade of this century (Alcaraz *et al.* 2009, Fernández *et al.* 2010). Yet, only recently did it become possible to design long-term monitoring programmes capable of providing ecosystem functioning information at the relevant spatial and temporal resolution for protected-area management. This capability is now possible after approximately two decades of data collection by MODIS. The Observatory at Doñana was devised in 2015, to address the need of assessing ecosystem functioning as a part of a broader effort to generate a pilot biodiversity monitoring infrastructure, supported by the European Strategy Forum on Research Infrastructures (ESFRI). This initiative was specifically designed to contribute to the establishment of the 'Lifewatch' pan-European technology infrastructure for biodiversity and ecosystem research (www.lifewatch.eu).

One of the main challenges we encountered when developing the observatory was translating the complex concept of ecosystem functioning into common language. Protected-area managers and decision-makers are usually familiar with the importance of monitoring water quality (e.g. in wetlands), endangered species populations, or the spread of invasive species, as examples of 'tangible' conservation-relevant targets. On the other hand, the value of monitoring the temporal dynamics of carbon and water fluxes is less often recognised by protected-area managers, and must come hand in hand with detailed explanations and interpretations of the indices. This complexity has been highlighted as one of the main problems when incorporating ecosystem-functioning variables into conservation biology (Cabello *et al.* 2012). To deal with this challenge we designed a set of indices and graphics to make complex information on ecosystem dynamics accesible to non-experts. However, this was and remains a challenge. As such, it requires further input from multidisciplinary teams, including end users.

The observatory was created as a 'proof of concept', a demonstrative action through which protected-area managers were able to understand and appreciate the utility of satellite remote sensing for the remote monitoring of ecosystem function. To advance this objective (alongside scientific papers, conferences, and workshops) we developed a prototype visual laboratory through which we made available the visualisations and graphics of the observatory. We selected a subset of categories that could be selected by the users to understand the stability of a certain ecosystem following the intense drought event of 2005. Once the user selected the category of interest, the system displayed the results along with an informative story line. As already noted by other authors (McInerny *et al.* 2014, Moreno *et al.* 2014b), these graphical and visualisation tools facilitate the creation of a framework that enabled a common understanding, and the informative storyline helps to guide users in the process of learning and engagement (Krzywinski and Cairo 2013).

Currently, this approach is being transferred to other ecosystems within the framework of Life AdaptaMed project (www .lifeadaptamed.eu/), which aims to promote the management of Mediterranean ecosystems, to mitigate the negative effects of climate change on key ecosystem services. Remote monitoring of ecosystem functioning is being used to analyse the stability of different ecosystems to management actions that are promoted to increase resilience. As such, our project could be seen as an important source of information for managers of other protected areas, in addition to Doñana National Park.

6.6 CONCLUSIONS

In this case study, we illustrated how we have developed a remote sensing-based observatory for monitoring ecosystem function in protected areas, based on Earth observation data. Our observatory facilitated the monitoring of the integrity of ecosystems that are highly vulnerable to climate change, as well as providing a synoptic overview of the results of restoration actions in terms of carbon and water balance at the ecosystem level. It can be used to deliver alerts, such as drought impacts on vegetation and its resilience. It has also demonstrated that the clear-cutting of pine forests is, as intended, alleviating the pressure over the aquifer in the short term.

The strength of our system lies in the simple visualisation tools, which aimed to translate complex issues into common language. The aim was to facilitate knowledge exchange among scientists, managers, and stakeholders. Such monitoring is expected to help practitioners not only to understand the effects of perturbations on an ecosystem's stability but also to promote adaptive management actions to revert or mitigate these effects.

The vulnerability of ecosystems to climate change is a worldwide issue, affecting a wide range of systems (Peñuelas *et al.* 2001, Granier *et al.* 2007, Soja *et al.* 2007, Harris *et al.* 2014). Therefore, the implementation of tools for understanding the vulnerability of ecosystems to climate change is a pressing issue for their conservation, and the benefits and goods that they provide (Brang 2001, Bodin and Wiman 2007). Moreover, as the Convention on Biological Diversity (CBD) has already expressed, there is an urgent need for managing ecosystems, for increasing resilience, and for assuring the preservation of biodiversity in protected areas. The tools and graphics developed in the Ecosystem Observatory of Doñana, which are all based on free satellite remote sensing data, are easy to transfer to other Mediterranean protected areas.

ACKNOWLEDGEMENTS

This study was funded by the Excellence Research Program of Junta de Andalucia, project no. RNM-6685, and by project 'Adaptación y mejora de la internacionalización de la e-infraestructura ICTS-RBD para la ESFRI-Lifewatch' project (ref: FICTS-2014/01/AIC-A-2011–0706). Paula Escribano was partly supported by funds from AdaptaMed project (LIFE CCA/ES/000612). We would like to thank Miguel Delibes and Mónica García for their help and support of the project and their valuable ideas, and María López-Rodríguez for the enriching discussions on science–management interfaces. We also thank the Instituto Geográfico Nacional for providing the orthophotos used in this chapter.

REFERENCES

Alcaraz-Segura, D., Cabello, J., Paruelo, J. M., and Delibes, M. (2009). Use of descriptors of ecosystem functioning for monitoring a national park network: a remote sensing approach. *Environmental Management*, **43**, 38–48.

Baettig, M. B., Wild, M., and Imboden, D. M. (2007). A climate change index: where climate change may be most prominent in the 21st century. *Geophysical Research Letters*, **34**, 1–6.

Baret, F. and Guyot, G. (1991). Potentials and limits of vegetation indices for LAI and APAR assessment. *Remote Sensing of Environment*, **35**, 161–173.

Bodin, P. and Wiman, B. L. (2007). The usefulness of stability concepts in forest management when coping with increasing climate uncertainties. *Forest Ecology and Management*, **242**, 541–552.

Brang, P. (2001). Resistance and elasticity: promising concepts for the management of protection forests in the European Alps. *Forest Ecology and Management*, **145**, 107–119.

Cabello, J., Fernández, N., Alcaraz-Segura, D., *et al.* (2012). The ecosystem functioning dimension in conservation: insights from remote sensing. *Biodiversity and Conservation*, **21**, 3287–3305.

Carnicer, J., Coll, M., Ninyerola, M., *et al.* (2011). Widespread crown condition decline, food web disruption, and amplified tree mortality with increased climate change-type drought. *Proceedings of the National Academy of Sciences*, **108**, 1474–1478.

Chapin, F. S., III, Matson, P. A., and Vitousek, P. (2011). *Principles of Terrestrial Ecosystem Ecology*. New York: Springer Science & Business Media.

Ciais, P., Reichstein, M., Viovy, N., *et al.* (2003). Europe-wide reduction in primary productivity caused by the heat and drought in 2003. *Nature*, **437**, 529–533.

Convention on Biological Diversity (2010). Decision X/2: the strategic plan for biodiversity 2011–2020 and the Aichi biodiversity targets. Conference of the Parties to the Convention on Biological Diversity, Tenth meeting, Nagoya, Japan, 18–29 October.

Costelloe, B., Collen, B., Milner-Gulland, E. J., *et al.* (2016). Global biodiversity indicators reflect the modeled impacts of protected area policy change. *Conservation Letters*, **9**, 14–20.

Cui, X., Gibbes, C., Southworth, J., and Waylen, P. (2013). Using remote sensing to quantify vegetation change and ecological resilience in a semi-arid system. *Land*, **2**, 108–130.

Dakos, V., Scheffer, M., van Nes, E. H., *et al.* (2008). Slowing down as an early warning signal for abrupt climate change. *Proceedings of the National Academy of Sciences*, **105**, 14308–14312.

De Keersmaecker, W., Lhermitte, S., Honnay, O., *et al.* (2014). How to measure ecosystem stability? An evaluation of the reliability of stability metrics based on remote sensing time series across the major global ecosystems. *Global Change Biology*, **20**, 2149–2161.

De la Riva, E. G., Lloret, F., Pérez-Ramos, I.M., *et al.* (2016). The importance of functional diversity in the stability of Mediterranean shrubland communities after the impact of extreme climatic events. *Journal of Plant Ecology*, **10**, 281–293.

Del Cacho, M. and Lloret, F. (2012). Resilience of Mediterranean shrubland to a severe drought episode: the role of seed bank and seedling emergence. *Plant Biology*, **14**, 458–466.

Escribano P., Fernández N., Oyonarte C., *et al.* (2017). Resistance and resilience metrics as tools for managers of protected areas: a remote sensing approach. The 24th MEDECOS and 13th AEET Meeting, Seville, Spain, 31 January–4 February.

Fernández, N., Paruelo, J. M., and Delibes, M. (2010). Ecosystem functioning of protected and altered Mediterranean environments: a remote sensing classification in Doñana, Spain. *Remote Sensing of Environment*, **114**, 211–220.

Fernández, N., Román, J., and Delibes, M. (2016). Variability in primary productivity determines metapopulation dynamics. *Proceedings of the Royal Society B*, **283**, 20152998.

Fisher, J. B. (2013). Land-atmosphere interactions: evapotranspiration. In Njoku, E., ed., *Encyclopedia of Remote Sensing*. Berlin: Springer-Verlag, pp. 1–5.

García, M., Fernández, N., Villagarcía, L., *et al.* (2014). Accuracy of the Temperature–Vegetation Dryness Index using MODIS under water-limited vs. energy-limited evapotranspiration conditions. *Remote Sensing of Environment*, **149**, 100–117.

Giorgi, F. (2006). Climate change hot-spots. *Geophysical Research Letters*, **33**, L08707.

Gómez-Rodríguez C., Díaz-Paniagua, C., and Bustamante, J. (2010). Evidence of hydroperiod shortening in a preserved system of temporary ponds. *Remote Sensing*, **2**, 1439–1462.

Gouveia, C. M., Trigo, R. M., Beguería, S., and Vicente-Serrano, S. M. (2017). Drought impacts on vegetation activity in the Mediterranean region: an assessment using remote sensing data and multi-scale drought indicators. *Global and Planetary Change*, **151**, 15–27.

Granier, A., Reichstein, M., Bréda, N., *et al.* (2007). Evidence for soil water control on carbon and water dynamics in European forests during the extremely dry year: 2003. *Agricultural and Forest Meteorology*, **143**, 123–145.

Green, A. J., Alcorlo, P., Peeters, E. T., *et al.* (2017). Creating a safe operating space for wetlands in a changing climate. *Frontiers in Ecology and the Environment*, **15**, 99–10.

Grimm, V. and Wissel, C. (1997). Babel, or the ecological stability discussions: an inventory and analysis of terminology and a guide for avoiding confusion. *Oecologia*, **109**, 323–334.

Harris, A., Carr, A. S., and Dash, J. (2014). Remote sensing of vegetation cover dynamics and resilience across southern Africa. *International Journal of Applied Earth Observation and Geoinformation*, **28**, 131–139.

Huang, L., He, B., Chen, A., *et al.* (2016). Drought dominates the interannual variability in global terrestrial net primary production by controlling semi-arid ecosystems. *Scientific Reports*, **6**, 24639.

Huete, A. (2016). Ecology: vegetation's responses to climate variability. *Nature*, **531**, 181–182.

Huete, A., Didan, K., Miura, T., *et al.* (2002). Overview of the radiometric and biophysical performance of the MODIS vegetation indices. *Remote Sensing of Environment*, **83**, 195–213.

IPCC (2012). *Managing the Risks of Extreme Events and Disasters to Advance Climate Change Adaptation. A Special Report of Working Groups I and II of the Intergovernmental Panel on Climate Change*. Cambridge: Cambridge University Press.

Ivits, E., Horion, S., Erhard, M., and Fensholt, R. (2016). Assessing European ecosystem stability to drought in the vegetation growing season. *Global Ecology and Biogeography*, **25**, 1131–1143.

Krzywinski, M. and Cairo, A. (2013). Points of view: storytelling. *Nature Methods*, **10**, 687.

Lindenmayer, D. B. and Likens, G. E. (2009). Adaptive monitoring: a new paradigm for long-term research and monitoring. *Trends in Ecology & Evolution*, **24**, 482–486.

Lloret, F., Keeling, E. G., and Sala, A. (2011). Components of tree resilience: effects of successive low-growth episodes in old ponderosa pine forests. *Oikos*, **120**, 1909–1920.

López-Rodríguez, M. D., Castro, H., Arenas, M., *et al.* (2017). Exploring institutional mechanisms for scientific input into the management cycle of the national protected area network of Peru: gaps and opportunities. *Environmental Management*, **60**, 1022–1041.

McInerny, G. J., Chen, M., Freeman, R., *et al.* (2014). Information visualisation for science and policy: engaging users and avoiding bias. *Trends in Ecology & Evolution*, **29**, 148–157.

Monteith, J. L. (1965), Evaporation and the environment. *Symposium of the Society of Exploratory Biology*, **19**, 205–234.

Moreno, Á., García-Haro, F. J., Martínez, B., and Gilabert, M. A. (2014a). Noise reduction and gap filling of fAPAR time series using an adapted local regression filter. *Remote Sensing*, **6**, 8238–8260.

Moreno, J., Palomo, I., Escalera, J., Martín-López, B., and Montes, C. (2014b). incorporating ecosystem services into ecosystem-based management to deal with complexity: a participative mental model approach. *Landscape Ecology*, **29**, 1407–1421.

Mumby, P. J. and Anthony, K. (2015). Resilience metrics to inform ecosystem management under global change with application to coral reefs. *Methods in Ecology and Evolution*, **6**, 1088–1096.

Nagendra, H., Lucas, R., Honrado, J. P., *et al.* (2013). Remote sensing for conservation monitoring: assessing protected areas, habitat extent, habitat condition, species diversity, and threats. *Ecological Indicators*, **33**, 45–59.

Peñuelas, J., Lloret, F., and Montoya, R. (2001). Severe drought effects on Mediterranean woody flora in Spain. *Forest Science*, **47**, 214–218.

Pérez-Ramos, I. M., Díaz-Delgado, R., Riva, E. G., *et al.* (2017). Climate variability and community stability in Mediterranean shrublands: the role of functional diversity and soil environment. *Journal of Ecology*, **105**, 1335–1346.

Pickup, G. and Foran, B. D. (1987). The use of spectral and spatial variability to monitor cover change on inert landscapes. *Remote Sensing of Environment*, **23**, 351–363.

R Core Team (2017). *R: A Language and Environment for Statistical Computing*. Vienna: R Foundation for Statistical Computing. See www.R-project.org.

Rendón, M. A., Green, A. J., Aguilera, E., and Almaraz, P. (2008). Status, distribution and long-term changes in the waterbird community wintering in Doñana, south–west Spain. *Biological Conservation*, **141**, 1371–1388.

Sandholt, I., Rasmussen, K., and Andersen, J. (2002). A simple interpretation of the surface temperature/vegetation index space for assessment of surface moisture status. *Remote Sensing of Environment*, **79**, 213–224.

Serrano, L. and Zunzunegui, M. (2008). The relevance of preserving temporary ponds during drought: hydrological and vegetation changes over a 16-year period in the Doñana National Park (south-west Spain). *Aquatic Conservation: Marine and Freshwater Ecosystems*, **18**, 261–279.

Simoniello, T., Lanfredi, M., Liberti, M., Coppola, R., and Macchiato, M. (2008). Estimation of vegetation cover resilience from satellite time series. *Hydrology and Earth System Sciences Discussions*, **5**, 511–546.

Soja, A. J., Tchebakova, N. M., French, N. H. F., *et al.* (2007), Climate induced boreal forest change: predictions versus current observations. *Global and Planetary Change*, **56**, 274–296.

Soriguer, R. C., Rodríguez-Sierra, A., and Domínguez-Nevado, L. (2001). Análisis de la incidencia de los grandes herbívoros en la marisma y vera del Parque Nacional de Doñana. Madrid: Organismo Autónomo de Parques Nacionales.

Van Ruijven J. and Berendse F. (2010). Diversity enhances community recovery, but not resistance, after drought. *Journal of Ecology*, **98**, 81–86.

Verbesselt, J., Hyndman, R., Newnham, G., and Culvenor, D. (2010). Detecting trend and seasonal changes in satellite image time series. *Remote Sensing of Environment*, **114**, 106–115.

Vogel, A., Scherer-Lorenzen, M., Weigelt, A., and Moen, J. (2012). Grassland resistance and resilience after drought depends on management intensity and species richness. *PLOS ONE*, **7**, 1–10.

Washington-Allen, R. A., Ramsey, R. D., West, N. E., and Norton, B. E. (2008). Quantification of the ecological resilience of drylands using digital remote sensing. *Ecology and Society*, **13**, 33.

Watson, J. E., Dudley, N., Segan, D. B., and Hockings, M. (2014). The performance and potential of protected areas. *Nature*, **515**, 67–73.

Worboys, G. L., Lockwood, M., Kothari, A., Feary, S., and Pulsford, I., eds. (2015). *Protected Area Governance and Management*. Canberra: ANU Press.

7 Predicting Mule Deer Harvests in Real Time

Integrating Satellite Remote Sensing Measures of Forage Quality and Climate in Idaho, United States

Mark Hebblewhite, Mark Hurley, Paul Lukacs, and Josh Nowak

7.1 INTRODUCTION

Recreational and subsistence hunting of ungulates is important to social and ecological processes worldwide. This is especially so for deer species. For example, in the United States, in 2015 an estimated 16 million recreational hunters harvested 2.8 million male white-tailed deer (*Odocoileus virginianus*) (Quality Deer Management Association 2017). In Europe, recreational ungulate harvests are a huge economic activity with over 1,000,000 roe deer (*Capreolus capreolus*) harvested per year in Germany alone (Apollonio *et al.* 2010a, 2010b). Recreational harvests are a billion-dollar industry across western countries (e.g. in North America, Europe, and New Zealand), and especially so in rural economies. Subsistence hunting occurs throughout the developing world, and bushmeat is often a critical source of protein for millions of humans (Brashares *et al.* 2004).

Ungulates are a key species in many ecosystems because they shape how the ecosystem works. This can be seen in their impact on tree regeneration and succession patterns (Cote *et al.* 2004), their influence on soil nitrogen and carbon cycling (Hobbs 1996, Gordon *et al.* 2004), and their role as prey for globally endangered large carnivores (Ripple *et al.* 2015). Across the globe, many ungulate species are threatened, due to population declines as a result of habitat loss

(Ripple *et al.* 2015). Yet in other regions, ungulates are locally over-abundant, causing ecosystem damage to plants, altering vegetation communities, and even causing other species to become endangered (Cote *et al.* 2004). Thus, whether regulating the overharvests on declining species, or increasing harvests on overabundant species, the management of ungulate harvests is a critical socioecological process.

Despite the global ecological and economic importance of ungulate harvests, the scientific basis for managing recreational and, especially, subsistence hunting is extremely variable. In Europe, there is great variability in implementing scientifically based ungulate harvest regulations, reflecting the variation in the quality of monitoring of ungulate population size, trends, and harvests (Apollonio *et al.* 2010a, 2010b). The history of ungulate conservation in North America is one of unregulated overexploitation, which led to near extinction of species. In developing countries, where bushmeat subsistence hunting trade dominates wildlife mortality and food supply, history seems destined to repeat itself, as there is often no regulation of harvests, leading to overhunting. Indeed, bushmeat hunting is one of the key factors for ungulate endangerment, especially in Africa and Southeast Asia (Ripple *et al.* 2015). Thus, around the world, improved management of ungulate harvests would improve the likelihood that species would not become threatened, which in turn would lead to better outcomes for people who rely on subsistence hunting and for those seeking recreational hunting opportunities.

The North American model of wildlife conservation provides an example of recreational ungulate harvests that is theoretically based on scientific management of ungulate populations and their habitat to support recreational harvests (Clark and Miloy 2014). Recreational hunters are taxed on guns and ammunition in the USA through the Federal Aid in Wildlife Restoration Act of 1937 (better known as the Pittman–Robertson Act), and pay licence fees that are used to support government-led scientific management of hunted species and their habitats. This 'user–pay' system generated US$1.1 billion in revenue under the Pittman–Robertson Act in 2016 alone (United States Fish and Wildlife Service 2016). Although this system has provided financial resources for wildlife management in the USA for decades, it is not perfect (Silvy 2012). A decline in the popularity of hunting has led to decreased revenue (Silvy 2012), and the scientific basis for harvest levels is often equivocal. Current conservation challenges, such as climate change and non-game wildlife and ecosystem protection, add layers of complication to managing hunted species.

Harvests in North America (and elsewhere) are generally goal-oriented, with an objective that could involve increasing, decreasing, or maintaining populations and their harvests to be as constant as possible (Sinclair *et al.* 2005). The goal may be to maximise the short-term harvest yield, for example in bushmeat hunting, or sustain a long-term harvest. Goals can also be oriented towards other management objectives. For instance, a goal could also be to help recover threatened species, for example caribou (*Rangifer tarandus*), thus necessitating the harvests to be legally closed for a period of time. Overabundant ungulates can also cause agricultural damage in the tens to hundreds of millions of US dollars in each US state (Silvy 2012). Achieving these different wildlife harvest and population goals is the field of wildlife management, which has been long-defined as the art and science of achieving management objectives (Leopold 1933, Sinclair *et al.* 2005). This perspective recognises the importance of the scientific basis of wildlife management together with the fact that wildlife management is a social process, linked to societally defined goals and objectives (Leopold 1933). Consequently, wildlife managers indirectly manage ungulate populations by setting harvest limits.

Wildlife managers are often faced with setting harvest quotas of ungulates, 1 or 2 years before harvest implementation, even in places such as North America that utilises scientifically based ungulate management. A lag occurs because of legislative and policy delays, changes to licensing regulations, and changing human hunting behaviour (harvest effort). These delays can, at the very least, cause inefficiencies and, at the worst, can themselves be the cause of population fluctuations that the harvest aims to avoid (Fryxell *et al.* 2010). For example, Fryxell *et al.* (2010) showed that even 1- or 2-year lags in changes in harvest regulations for white-tailed deer in Canada and Eurasian elk (moose; *Alces alces*) in Norway led to long-term oscillations and cycles in ungulate population size that destabilised both populations and harvests. Moreover, because ungulates play a key role in ecosystems, both as herbivores and as prey for large carnivores, such hunting-induced fluctuations in population size could also have cascading effects, through food webs, to ecosystem productivity (Hobbs 1996, Cote *et al.* 2004). Thus, wildlife managers would benefit from a method to predict population size in near real time, and thus be able to adjust harvests, to optimise harvest levels for specific goals and also to minimise unintended population fluctuations. This is where the growing appreciation for the ability of satellite remote sensing tools to successfully reflect ecosystem dynamics for ungulates comes into play.

Here, we illustrate a predictive harvest management model, based on satellite remote sensing and ground-based inventory data, and apply it to improve the harvest of mule deer (*Odocoileus hemionus*, Figure 7.1), an economically and ecologically important ungulate across western North America. We first introduce our study area in Idaho, a state in northwestern USA, and then provide some general background on mule deer ecology and harvest management across western North America. Next, we review the development of a Bayesian integrated population model (IPM) for mule deer in Idaho across 11 population units and 11 years of data (Hurley 2016). We then illustrate how satellite remote sensing data improved the ability to predict overwinter mule deer fawn survival, a key demographic parameter. This information enabled managers to set harvests in advance of the end of the biological year for mule deer (December) for the following harvest season (November of the following year). We then introduce a system that utilises real-time remote sensing information on environmental and climate conditions to improve ungulate harvest management. This is illustrated as a case study on two deer

FIGURE 7.1 Male and female mule deer during October.

populations. Finally, because wildlife management is a unique blend of natural and social science, we provide a narrative of our implementation process that combined remote sensing science, traditional wildlife biology, and software development (Nowak *et al.* 2017) to deliver this new system in a wildlife management agency, Idaho Department of Fish and Game (IDFG).

7.2 STUDY AREA

Our study area covered a wide climatic, predation, and habitat gradient of mule deer range in Idaho (Figure 7.2). We monitored overwinter survival of fawns in 11 population management units (PMUs) across central and southern Idaho, which were themselves comprised of 28 different game management units (GMUs). In Idaho, GMUs are nested within PMUs, which are grouped together to represent ecological (interbreeding) populations, and which form the basis for management. Mule deer PMUs were divided into three ecotypes: coniferous forests, shrub-steppe, and aspen woodlands (Hurley *et al.* 2014, 2017). Elevation

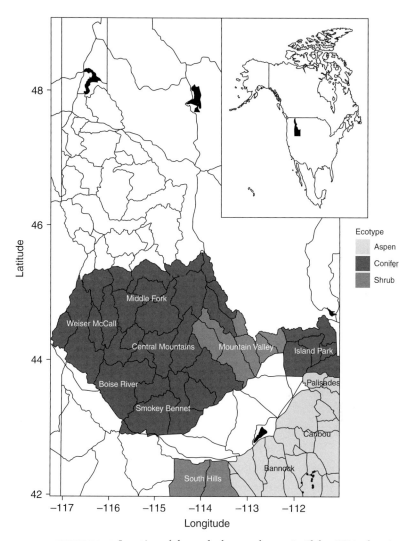

FIGURE 7.2 Location of the mule deer study area in Idaho, USA, showing the 11 PMUs in three different ecotypes of central and southern Idaho.

and topographic gradients within PMUs affect snow depths and temperature in winter, and precipitation and growing-season length in the summer, with elevation increasing from the southwest to the northeast. For more details on the study area, see Hurley *et al.* (2014, 2017).

7.3   MULE DEER ECOLOGY AND MANAGEMENT

Mule deer are medium-sized ungulates, occurring in the desert, semi-arid, and temperate ecosystems of western North America. Mule deer adult males weigh c. 100 kg, and females, c. 70 kg. They are browsers and mixed feeders, with their diet dominated by woody plants in the winter and nutritionally rich herbaceous vegetation in spring, summer, and autumn. Mule deer are polygynous breeders, with males competing for breeding opportunities during the November rutting season. For the rest of the year they are gregarious, living in matrilineal mixed female groups and separate male bachelor groups. Mule deer give birth to one to three fawns (i.e. juveniles) in late May and early June (Figure 7.3), the number of which is driven by the nutritional condition of the mother. Neonatal fawns hide for the first month of their lives, during lactation, and then slowly wean by late summer, and are capable of breeding, usually as yearlings, but often as subadults (2.5 years old).

Mule deer population dynamics are primarily driven by the variability in overwinter survival of juveniles from 6 to 12 months of age, but also by female survival (Unsworth *et al.* 1999, Bishop *et al.* 2009, Hurley *et al.* 2011). As with many other ungulates, female fecundity (e.g. pregnancy rate) is often high and constant (Gaillard *et al.* 2000). Due to their polygynous breeding system, male mortality has a minimal effect on population dynamics. Overwinter survival is driven primarily by summer-growing-season conditions, which influence nutrition, and, secondarily, by winter severity, which drains nutritional stores (Hurley *et al.* 2014). The overwinter survival of mule deer is sensitive to snow depth, snow quality, and winter temperatures, which deplete stored fat reserves. The latter are predicted by exposure to high-quality forage over the preceding spring and autumn growing seasons. This is particularly the case in semi-arid systems, where autumn nutrition is often driven by a second growing season, after the mid-summer drought (Hurley *et al.* 2014, Monteith *et al.* 2014).

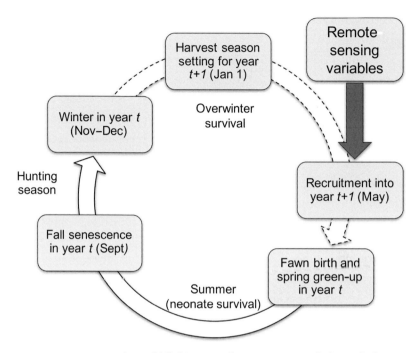

FIGURE 7.3 Annual life-history and management cycle for mule deer fawns in year *t* from birth, through summer and winter survival to recruitment into the population in relation to the management setting of mule deer harvest in year *t* + 1 in Idaho, USA. Winter begins in November or December and winter survival is measured from 15 December to June 1 (dashed line). Season setting for harvest management begins in early January and final seasons are set in early March of each year. Remote sensing variables available in December could predict overwinter survival and recruitment before harvest setting in January.

Mule deer have a similar life-history to many other medium-sized ungulates (Figure 7.3), and provide a model system for exploring how information that is derived from remotely sensed weather and habitat-related variables can be used to understand and model mule deer population dynamics. Variables derived from remote sensing correlate with the body mass of 6-month-old mule deer fawns entering winter, reflecting the links between growing-season weather, plant growth, and deer body-fat stores (Figure 7.3). Hurley *et al.* (2014) showed that, in Idaho, Moderate Resolution Imaging

Spectroradiometer (MODIS) data successfully predicted overwinter mule deer fawn survival.

Mule deer are also a key species for harvest by recreational hunters across western North America. In 2015 in Idaho, around 200,000 hunters harvested almost 60,000 mule deer (Idaho Department of Fish and Game 2013). The overall revenue generated by hunting is difficult to quantify, but direct revenue from fees from resident hunters (i.e. hunters living in Idaho) is more than US$1,100,000 in Idaho alone. Typically, state agencies charge non-resident hunters (i.e. outside of Idaho) up to five to ten times more for hunting permits, generating additional direct hunting revenues for state governments. In addition, indirect economic benefits (e.g. gasoline, equipment, hotel, lodging) can be 10–20 times greater than direct fee revenue. Although there is a preference in European and North American hunting traditions for trophy males, including mule deer, this form of selective harvesting has been shown to have limited consequences for population size and fluctuations. Hunting seasons are almost exclusively during the autumn (September–December, Figure 7.3). Deer harvests are managed at the scale of the spatial management units (GMUs within the PMUs described above), with hunting regulations varying between PMUs. This variation between PMUs results in a mix of open seasons (where either sex of deer can be harvested anywhere), quota systems (e.g. limited entry hunt), or more restrictive hunting seasons (lottery) where trophy quality (antler size) is a primary focus. Population information is collected by state wildlife agencies from the GMUs and PMUs.

A key challenge for improving the management of harvests is the need to set hunting levels before the end of a biological year (Figure 7.3). From a biological viewpoint, it would be ideal to wait until a biological year is completed in June, collect information about mule deer population dynamics and harvest success, and then set the harvest season for the immediate autumn hunting season. But administration introduces delays and constraints on the ability of wildlife managers to make rapid changes to harvests in response to changing population dynamics. For example, Fish and Game Commissions govern changes in harvest regulations in all US states, and they review all proposed harvests each year. Subsequent feedback from the public can see decisions changed. Once approved, these changes must make their way into printed regulations, lotteries for coveted restricted hunting permits, and business systems that sell hunting permits to the public online. Each of these steps slows down the ability to make rapid changes in harvests. But the administration of hunting means that many US states set hunting season changes in midwinter, immediately following the hunting season, when only harvest statistics are known, for example in Idaho in early January. Biological information about the population dynamics are usually unknown at the time of the process of changing hunting seasons (season setting, usually conducted in midwinter), creating a mismatch between administrative and biological systems. If wildlife managers had a way to predict population trends over the winter at the time of season setting, in December or January, this would potentially resolve this mismatch.

7.4 MODELLING MULE DEER POPULATION DYNAMICS

Wildlife managers have used a variety of methods since the 1930s to assess and model ungulate populations. These methods include indirect indices such as pellet counts, ground counts, and aerial counts; population reconstruction from harvest data; and population modelling from individually marked animals (Williams *et al.* 2002, Silvy 2012). Historically, wildlife managers and scientists have spent a lot of time and effort comparing and contrasting methods in search of the 'optimal' method. In the last decade, many practitioners now try to combine information from numerous sources, and have come to appreciate that relying on any one particular method rarely is the best avenue. Increasingly, the best tools to integrate different data types into models of wildlife population dynamics are Bayesian IPMs.

Bayesian IPM for Mule Deer

Here, we briefly describe the development of an IPM for mule deer in Idaho, which can combine multiple sources of population information and harvest statistics, and can account for imperfect detection processes to model mule deer population dynamics. Full details of the IPM are reported in several previous publications (Hurley *et al.* 2014, Hurley 2016, Hurley *et al.* 2017). IPMs allow for the incorporation of many different types of population data of varying quality. These data are integrated into a population model to provide improved rigour in both population estimates and projections (Besbeas *et al.* 2002, Johnson *et al.* 2010, Schaub and Kery 2012). We developed a Bayesian state-space IPM (Besbeas *et al.* 2002, Kery and Schaub 2012) to forecast population dynamics from the combined estimates of vital rates, harvests, and, when available, abundance data.

We used a post-breeding, sex-specific, and age-structured matrix model (Caswell 2000), modified from a standard large herbivore model in the timing of inclusion of fawns. Ecological parameters are: fawn to adult female ratio, adult male to adult female ratio, harvest data, population size, and measures of sex-specific survival of fawns and adults. Overwinter survival of fawns is the most important parameter affecting population trends. To obtain estimates, the IDFG conducted aerial PMU-wide population surveys using sightability models, corrected for visibility bias once every 4–5 years (Unsworth *et al.* 1994). Because of the high cost of conducting aerial surveys in general, aerial surveys were collected, on average, every 5 years. Early winter fawn to adult female ratios are a measure of age-specific fecundity and fawn survival for the first 6 months of life. The IDFG used helicopter surveys to estimate age and sex composition for each PMU every December (Unsworth *et al.* 1994). Adult male to adult female ratios are obtained concurrently with fawn to adult female ratios, and provide an estimate of the male (harvestable) portion of the population. The varying data collection methods between different data types (i.e. survival, population counts, etc.) mean that these ecological parameters have inherent differences in completeness (i.e. differences in missing years between PMUs) and a variance in different vital rates, for example, between years. Fortunately, Bayesian methods allow the integration of data of varying quality (Kery and Schaub 2012).

In our case study, we monitored the survival of 1,961 fawns (6-month-old males and females) and 1,061 adult females (monitored for an average of 4.5 years each) within six PMUs in Idaho between 2001 and 2013. Animal-capture protocols were approved by the Animal Care and Use Committee, Idaho Department of Fish and Game Wildlife Health Laboratory, Caldwell, Idaho, USA, and the University of Montana IACUC (protocol #02-11MHCFC-031811). Mortality was monitored from ground-based telemetry at least once a month between capture (7 December to 15 January) and 1 June. If radio signals could not be detected from the ground within 1 week, animals were located via aircraft. When a mortality was detected, the cause of death was determined using a standard protocol (see Hurley *et al.* (2014) and Hurley (2016) for additional details). Despite not having as much information in other settings, the same kind of IPM framework can be developed with reduced data (e.g. just aerial counts and information about the harvest, e.g. Ahrestani *et al.* 2013) to improve harvests of ungulates.

Finally, we used harvest data to estimate the adult male to adult female ratios. To estimate the total number of deer in the population at 15 December each year, the harvest is subtracted from each age class and then the current number of 6-month-old fawns (N_y) was added to the estimate. Idaho estimated harvests by requiring hunters to report hunting success; for those who do not report, officials follow up with a telephone survey (Idaho Department of Fish and Game 2013). The survey accounted for the biased reporting rate, based on hunting success and hunter demographics (Idaho Department of Fish and Game 2013). Harvests are specific to antlered versus antlerless mule deer and the number of antler points on the male deer.

7.5 INTEGRATION OF MODIS REMOTE SENSING VARIABLES INTO POPULATION MODELS

The power of remote sensing tools to monitor ecosystem dynamics at a variety of spatio-temporal scales has been underutilised to date by wildlife management agencies (Silvy 2012). Decades of applied research have demonstrated the efficacy of using satellite remote sensing tools to explain ecological dynamics, including ungulate ecology, behaviour, movement, resource selection, and even harvest and population dynamics. In the last decade, hundreds of published studies have demonstrated the potential of the Normalised Difference Vegetation Index (NDVI; see Chapter 2) in understanding terrestrial ecology (Pettorelli 2013). To briefly recap, NDVI is a function of the visible and near-infrared bands of the electromagnetic spectrum that reliably measures the amount of live, green vegetation. The NDVI can be measured by satellite sensors with different spatial resolutions, from low-resolution Advanced Very High Resolution Radiometer (AVHRR) data, to medium-resolution Moderate Resolution Imaging Spectroradiometer (MODIS) and Landsat data, to very-high-resolution data, captured by drones and hand-held sensors (Pettorelli 2013). In one of the first applications of MODIS NDVI data to ungulate ecology, we demonstrated that NDVI could predict the body condition, pregnancy, and dynamics of migrant and resident North American elk (*Cervus canadensis*) in Banff National Park in Alberta, Canada (Hebblewhite *et al.* 2008). Despite these research insights, there have been no examples that we know of in which remote sensing variables have been used to actively manage ungulate populations or harvest.

Here, we describe methods of using MODIS observations to derive summer plant productivity and winter snow conditions (snow cover). These covariates are then used to assess the effects of habitat and weather on winter survival of fawn mule deer at the GMU and PMU spatial scale. Remotely sensed data are spatially explicit and generally available to wildlife managers with a shorter delay (e.g.

weeks with MODIS) than *in situ* meteorological data (e.g. PRISM climate data available on a monthly to annual basis, http://prism .oregonstate.edu). This allows for the rapid integration of these data into harvest management.

In the first phase of analysis, we initially explored the relationship between overwinter mule deer survival and NDVI using 1-km resolution, 7-day composite AVHRR NDVI data obtained from the NOAA-14, -16, and -17 AVHRR sensor (maintained by the US Geological Survey; http://phenology.cr.usgs.gov/index.php; Eldenshink 2006). Initial survival models developed with AVHRR NDVI data proved effective in the development of a predictive model for overwinter mule deer survival (Hurley *et al.* 2014, 2017). In our second phase for this analysis, we switched to medium-resolution 250-m MODIS data, anticipating it would provide an even better explanation of mule deer survival. We estimated NDVI values from 16-day composite MODIS data obtained from the MOD13Q1 product, downloaded from the National Aeronautics and Space Administration (NASA) Land Products Distributed Active Archive Center (LP DAAC; https://lpdaac.usgs.gov/; Huete *et al.* 2002).

An additional advantage of using the medium-resolution MODIS data is that we were able to extract NDVI values only within open-canopied, non-forested landcover types, which are the types of habitats that mule deer use for foraging (Borowik *et al.* 2013). AVHRR data are too coarse in resolution to be able to do this at a scale relevant to mule deer. We extracted NDVI values from these vegetation types, filtering out those areas which had not burned within the preceding five years (Hurley *et al.* 2017). Burned and non-burned areas were determined using SAGEMAP, a Landsat-derived landcover model developed by the USGS specifically for the sage-brush steppe ecosystems of Idaho (https://sagemap.wr.usgs.gov). We restricted NDVI data to 15 March to 15 November to encompass the entire growing season for each mule deer population year, and excluded anomalies caused by varying snow conditions (i.e. June snowstorms). See Hurley *et al.* (2014, 2017) for more details about image processing for NDVI datasets.

NDVI data are time-series of growing season dynamics. Previous studies have extracted phenological parameters, characterising, for example, the start and peak of season, length of season, etc., as well as other metrics of phenological dynamics (Zhang *et al.* 2003, White *et al.* 2009). However, many of these parameters are understandably correlated, and challenging to model in a standard regression framework that assumes independence between independent covariates. This means that the development of a single, statistically independent metric to use in predictive models of overwinter mule deer survival can be difficult. Moreover, growing seasons can be complex, with multiple peaks. In semi-arid systems found in Idaho, there is often a second growing-season peak in the early autumn after a summer plant dormancy period because of the lack of precipitation during the July and August period. To reduce the complexity of analysis, we used a functional analysis (Ramsay and Silverman 2005) to quantitatively measure the shapes of the growing-season curves derived from either AVHRR NDVI or MODIS NDVI, for each population-year according to the methods of Hurley *et al.* (2014). This analysis is based on a multivariate functional analysis, which is similar to principal component analysis (PCA), or then functional PCA (FPCA) (see Hurley *et al.* 2014 for details). Additionally, it also accounts for the temporal autocorrelation within a time-series. In Idaho, we used functional analysis to identify the first two principal component scores of the NDVI phenology curves for each PMU. Together, these two PCA axes accounted for 74 per cent of the variance in phenology and could be interpreted as a measure of autumn growth (growth after the peak NDVI date) and spring green-up (growth at the start of the growing season). Hurley *et al.* (2014) provide more details, including FPCA R code (R Core Team 2014).

To measure winter snow cover, which is critical for determining energy expenditure and limiting access to forage, we used the MODIS snow-cover product (MOD10A2; Hall *et al.* 2002). The MODIS snow data product measures complete snow coverage in 8-day composites at a 250-m resolution. We highlight results of using MODIS snow cover during three winter periods, early winter (November–December), winter (January–March), and late winter (April), in two different ways. First, we estimated the percentage of each winter period that each pixel was covered by snow. This was a measure of fractional snow cover. Second, we used the number of weeks that more than 90 per cent of the PMU was covered by snow. Thus, the final variables for overwinter mule deer survival models that were derived from remote sensing data (both the snow and NDVI) were: functional analysis principal components for autumn (fall; FPC), functional analysis principal components for spring (SPC), mean percentage snow cover in November and December (ND%snow), mean percentage snow cover in January to March (W%snow), mean percentage snow cover in April (A%snow), number of weeks with > 90 per cent snow cover in November and December (FWeeks), number of weeks with >90 per cent snow cover in January to March (WWeeks), and number of weeks with >90 per cent snow cover in April (AWeeks). None of these seven covariates had a correlation coefficient that was greater than 0.7 (Hurley et al. 2017).

Overwinter fawn survival from 16 December to 1 June was then modelled in Bayesian hierarchical survival models (Royle and Dorazio 2006, Kery and Schaub 2012). We included the MODIS covariates at the appropriate spatial and temporal resolution at the nested GMU and then PMU scale (Hurley et al. 2014, 2017). The main goal was to develop the best predictive model of overwinter mule deer survival using MODIS covariates only available up to late December. We conducted model selection and tested the predictive capacity of survival models by relying on out-of-sample predictive performance (Hobbs and Hooten 2015), by testing the model's predictive ability against two withheld winters (2007, 2008, $n = 408$ individual mule deer). We compared predicted survival rates from each Bayesian survival model against observed survival rates within each year and PMU. Observed survival rates were estimated using the simple non-parametric, non-distributional Kaplan–Meier (K–M) estimator (Kaplan and Meier 1958). We used Pearson's correlation coefficient between observed and predicted survival to estimate the precision (R^2). We also considered a suite of random effect structures at different hierarchical levels, allowing the effects of MODIS covariates to vary across different PMUs; again, we direct readers interested in more details to Hurley *et al.* (2014, 2017).

Effects of MODIS Covariates on Mule Deer Survival

The body mass of 6-month-old fawns was strongly correlated to measurements made from AVHRR NDVI of both spring and autumn phenology across Idaho (Hurley *et al.* 2014). Earlier springs and later autumn growing seasons both increased overwinter survival of fawns (Hurley *et al.* 2014). Cumulative precipitation during winter also negatively affected mule deer survival, but NDVI variables had a higher predictive ability. The effect of the NDVI in autumn (FPCA component 1) on the autumn body mass of fawns (standardised beta coefficient (b) = 0.694, standard error (SE) = 0.209) was marginally greater than the effect of the NDVI in spring (FPCA component 2; standardised b = 0.652, SE = 0.206). This suggested that autumn conditions were of greater importance to the body development of fawns than the spring conditions in which they were born. With this first step demonstrating support of the AVHRR NDVI values affecting mule deer demography, we then applied the same FPCA methods to the MODIS NDVI data as described above, and then used these same simple metrics in regional predictive models of overwinter survival.

The most parsimonious survival model, with the highest external predictive power (R^2 = 0.704, Figure 7.4), was a simple function of winter per cent snow cover, autumn NDVI, and the number of weeks with complete snow cover in November and December, measured from MODIS data (Hurley *et al.* 2017). The most supported early prediction model (using only covariates before 1 January) included three covariates with a random PMU-level coefficient for autumn weeks with > 90 per cent snow cover with high explanatory capacity (R^2 = 0.818), but slightly lower external predictive power (R^2 = 0.590). Figure 7.4 shows the predictive power of the most parsimonious

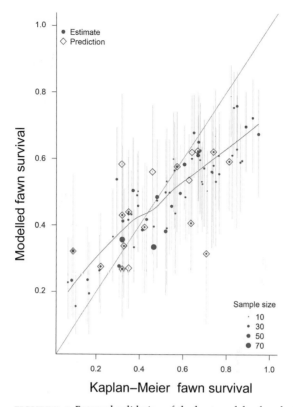

FIGURE 7.4 External validation of the best models of mule deer fawn survival in Idaho, 2003–2013. We conducted external validation by withholding survival data collected on 403 mule deer fawns in years 2007–2008. We used the models to predict survival of fawns and then compared the estimate to observed survival (Kaplan–Meier estimate) in the same study areas. This simple model successfully predicted survival using the MODIS remote sensing-derived variables of percentage of autumn weeks covered by snow, percentage of winter period covered by snow, and autumn NDVI growing season dynamics. The dark grey line is a spline fit to illustrate the bias of modelled survival estimates from observed estimates.

overwinter fawn mule deer survival model. These predictive models were integrated into our IPM to allow prediction of mule deer population dynamics for the following $(t + 1)$ autumn harvest season setting, in January 1, of year t. In the next section, we illustrate application of this to two PMU case-study areas.

7.6 CASE STUDY: PREDICTING POPULATION SIZE WITH REMOTE SENSING VARIABLES

We selected two contrasting PMUs – Boise River and Bannock – as case studies to illustrate the application of our MODIS-informed predictive population models. The Boise River PMU is generally poorer foraging habitat, more arid, and has lower-quality summer ranges than the more montane Bannock PMU. It is in south-central Idaho, is a semi-arid system, characterised by lower-elevation sagebrush-grass communities, steep canyons, and higher-elevation mesic summer ranges. In contrast, the Bannock PMU is in the Aspen ecoregion of Idaho, and is characterised by semi-arid lower-elevation winter ranges but higher productivity, moister and more productive summer ranges with deciduous Aspen forest cover, and productive understorey graminoid and forb communities. Mule deer in the Bannock PMU are usually in better body condition, with higher fawn growth rates (an indicator of habitat quality), averaging 10 per cent heavier in the autumn than those in the Boise PMU (Shallow *et al.* 2015). For more details, see Hurley *et al.* (2014, 2017).

Between 1998 and 2016, the IDFG monitored a median of 34 fawns per year (range 20–51) and a median of 47 adult females in the Boise River PMU. Over the same time, they monitored a median of 39 fawns per year (range 24–65) and a median of 53 adult females in the Bannock PMU. The annual fawn to adult female ratio and harvest statistics were also collected between 2000 and 2017. During the survey period there were five aerial counts of deer in the Boise River PMU and one aerial count in the Bannock PMU. These data were combined in our Bayesian IPM to estimate trends in population size of mule deer in each unit (Figure 7.5). Here, we focus on the application of the use of MODIS data on the percentage of autumn weeks covered by snow, average winter snow cover, and autumn plant phenology, summarised by our FPCA autumn phenology metric. For full details about the IPM model fitting, covariate effects, and model diagnostics in each PMU, see Hurley (2016) and Hurley *et al.* (2017).

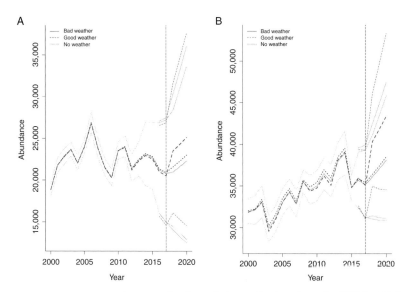

FIGURE 7.5 An example of population projections from the IPM for (A) the Boise River and (B) the Bannock PMUs in Idaho, USA. Shown are the predicted mule deer population size based on a Bayesian IPM from field data on mule deer survival, collected from 2000 to 2017, with 95% credible intervals. In the winter of 2016/2017, Idaho wildlife managers used the IPM, combined with remote sensing data obtained in autumn of 2016 regarding the number of autumn weeks covered by snow and autumn growing season phenology, to predict potential mule deer population size in the subsequent year, autumn 2017, for harvest, and two subsequent years (2017–2020). Three scenarios were considered; average weather conditions (i.e. 'no' weather), severe weather conditions, and mild weather conditions.

Figure 7.5 shows the mule deer population size and 95% Bayesian credible intervals from 2000 to 2017 from the IPM for the Boise (Figure 7.5A) and Bannock PMUs (Figure 7.5B). We used this model to test three scenarios of future conditions to inform possible harvest levels for autumn 2017 as well as the next 2 years (2018, 2019, ending in 2020). The three scenarios were the 'median' (i.e. average weather) scenario, a 'severe' winter (with a high percentage of autumn weeks with snow cover based on the 75th percentile), as well as a 'mild' winter (with a low percentage of autumn weeks with snow cover based on the 25th percentile). The Boise River PMU population

(Figure 7.5A) was dramatically affected by wide weather variation, while the Bannock PMU population (Figure 7.5B) was more strongly regulated by its own density-dependence in vital rates (for more details, see Hurley 2016). The population here was affected by weather, with much narrower ranges of predictions under the three weather scenarios.

In the Boise River PMU, under 'median' conditions, the model did not predict a change in the harvest based on population size in 2017. However, inclusion of MODIS data in autumn 2017, which constituted 'severe' weather conditions for mule deer based on our experience, managers would expect a significantly reduced population size and, thus, harvest potential. In contrast, if autumn 2017 was a 'mild' autumn with lower and later snow, and better growing conditions, the harvest in 2018 could be substantially higher (Figure 7.5A). In comparison to the Boise River PMU, predictions in the Bannock PMU were not as dramatically different under 'median' conditions compared to 'severe' and 'mild' autumn weather in 2017 for mule deer overwinter survival. Nonetheless, information from MODIS about the conditions of the autumn could obviously reduce the risk of overharvesting if, instead of an average year, the autumn was 'severe' in terms of snow cover and/or growing season phenology (Figure 7.5B).

IDFG wildlife managers can use this information to inform the prudent harvest of mule deer males and females in autumn hunting seasons. The actual harvest decision is obviously also a function of risk tolerance, stated management objectives for each PMU, and the IDFG wildlife managers' view of the likelihood of mild, average, or severe winters for mule deer. Regardless of the challenge of predicting future weather and the risk tolerance of wildlife managers, the consequences of risk-averse or risk-prone mule deer harvest management scenarios can be quantitatively assessed.

For example, in a population that is recovering from a recent population decline, wildlife managers might adopt a more risk-averse strategy, using our models to guide population recovery to a larger overall population size, and, hence, higher harvest in the long-run. The Boise River PMU provides an example of this scenario, whereby under both average and harsh future weather, the mule deer population was likely to continue to decline (Figure 7.5A). In contrast, for a population which is thought to be potentially close to its ecological carrying capacity, wildlife managers may adopt a more risk-tolerant harvest strategy, such as in the Bannock PMU. In the latter, under both average and mild weather conditions, the population was likely to increase. This suggests that a higher harvest may be possible in this population (Figure 7.5B). But our ability to make projections of mule deer population size for the next year is crucially dependent on the established empirical relationships between three key remote sensing variables derived from MODIS data, and illustrate the power of predictive modelling based on remote sensing for informing wildlife harvest management.

7.7 IMPLEMENTATION: FROM MODELS TO MANAGEMENT

We believe wildlife management agencies are currently underutilising the benefits of remote sensing data to help improve management of their wildlife populations and, importantly, to improve harvest management. In the most recent seventh edition of the *Wildlife Societies Techniques Manual*, for example, there is a dearth of remote sensing applications (Silvy 2012). This is despite hundreds of studies demonstrating the ability of remote sensing products to help explain ungulate (and other wildlife) population trends, spatial distribution, migration, and even harvest (Pettorelli 2013). We believe this important gap is because of missing crucial steps in the implementation and technology transfer of remote sensing science, in a manner that is accessible by practising wildlife managers. Here, we describe the practical steps we undertook, and the preconditions we believe were necessary for successful implementation of remotely sensed data into our mule deer population modelling case study.

We suggest that one of the most important prerequisites for incorporating remote sensing into harvest management is agency culture, in particular the relationship between research and management within a particular wildlife agency. This relationship varies across wildlife agencies around the world, from that of a service role, where research is directed to provide answers only to questions that wildlife managers envision, to a more isolated 'silo' approach, where researchers are often completely insulated from those who are implementing wildlife management objectives. Neither extreme provides ideal conditions for successful integration of research into management. We believe that an important precondition is good communication and collaboration between research and management agencies so that the research is both responsive to the needs of wildlife managers, but also has the flexibility to investigate visionary and creative solutions. Innovation is where healthy collaboration with university researchers can play a supporting, but not necessarily leading, role.

The IDFG has a long history of a visionary leadership approach in developing wildlife research that can be applied by harvest managers. Intra-agency support for research has been strong in Idaho for decades, as exemplified by the historical development of some of the first 'sightability' population models that were directly incorporated into management. Aerial surveys provide counts of ungulate species, which are affected by imperfect detection, whereby an unknown portion of the population is actually observed. This is crucial for harvest management because imperfect detection means that populations are most often being underestimated, and thus underharvested. Appreciating this challenge in the late 1970s, the IDFG directed research in collaboration with universities to develop rigorous statistical detectability estimators, which provided improved population estimates of ungulate species. The detectability model's generality and precision was then tested in field trials under different conditions (Unsworth et al. 1990, Hurley 1992). These resulted in a series of scientific papers (Samuel et al. 1987, 1992, Unsworth et al. 1994). However, scientific papers were not enough to translate science into management. One of the first computer software packages for wildlife management was developed by the IDFG and the University of Idaho in the 1990s for regional wildlife biologists to use and implement (Unsworth *et al.* 1994).

The relationships between wildlife managers and researchers paved the way in the IDFG for what we believe to be a second precondition for effective development and integration of remote sensing applications in wildlife management agencies: a centralised database structure. A large part of the success of our approach in using remote sensing products is dependent on centralised agency databases, where all regional aerial survey and other monitoring data for ungulate populations are housed, along with centralised harvest management databases. This is in contrast to many wildlife agencies where regional biologists are responsible for managing wildlife inventory data on specific local populations. Often, there is resistance to centralising data, and hence decision-making power, at headquarters. In other cases, the process of centralising the database structure highlights regional discrepancies in data collection protocols, datasets, metadata format, and even storage issues. Indeed, centralisation often exposes weaknesses in the science underlying wildlife management. All of these can be viewed as a hindrance and impediment to the centralisation of data. There are benefits to local data management since the data is best known locally and centralised databases can be expensive to manage. Yet, from our experiences in Idaho, we feel that a centralisation process, while it can be painful, leads to improvements in data collection protocols and is needed to scale up local ecological knowledge to management scales, and to be able to harness the capability of remote sensing to enhance wildlife management.

For us, the third step in the successful development of remote sensing applications, which resulted from this centralised database structure and research-friendly management culture, is the rigorous and regional monitoring of wildlife population trends. The IDFG developed a rigorous and state-wide monitoring programme for mule deer (and other species, including the North American elk) that involved standardised aerial survey methods and protocols, age and sex ratio data collection in a similar standardised protocol, and detailed population monitoring of individual animals, using radiotelemetry across representative parts of Idaho. This resulted in the tracking of thousands of individual mule deer across more than 11 monitoring units over a dozen years. It is these data that formed the foundations of the analysis in the study presented here. Even without such detailed individual demographic monitoring, aerial survey data and age and sex ratio data have been similarly shown to be highly correlated with remotely sensed data such as NDVI (Griffin *et al.* 2011, Brodie *et al.* 2013, Lukacs *et al.* unpublished work). Thus, our approach of using MODIS covariates to enhance harvest management will have widespread applications in other settings and species, for example in situations without data from individually marked animals and more typical aerial count and recruitment data.

Centralised data allowed the IDFG to identify PMUs that reflected biological mule deer populations. This use of data in the identification of management units is our fourth precondition. Often, wildlife management units that are used for harvest management are defined administratively (e.g. counties, voting districts) and are unrelated to biological populations. This mismatch between units used for harvest management and the biological units of interbreeding populations is often historical, and a common problem across state wildlife management in the USA. The scaling up of these game management units (GMUs) to the larger-scale PMUs preserved the historical structure of GMUs for harvest management, while recognising the biological importance of understanding management at the true biological population level. It is only through this population-based, rigorous statistical monitoring that we were able to develop the regional-scale survival and integrated population models.

The next step in our process was collaborative workshops, presentations, and internal agency communication to disseminate the results of the Idaho research biologists to managers within the IDFG. Having close, regular communication channels between researchers and managers was vital to demonstrating the success of remote sensing applications and for receiving feedback and suggestions from managers about how they could imagine using such technological advances. In other words, the advice and frequent communication from managers to researchers on how to actually make the research useful was critical. Universities are often too isolated from wildlife managers. We feel that strong academic research, agency research, and agency-research–management communication was critical to the success of our approach. Also important to our success was the role of the Montana Cooperative Wildlife Research Unit in facilitating cooperative workshops amongst ungulate managers from across the western USA. In our particular case, it was by chance this was facilitated by this unit, located at the University of Montana (east of Idaho), but it is the mission of the Cooperative Wildlife Research Unit system in the United States Geological Survey to facilitate exactly these kinds of collaborative approaches across all states in the USA. Inter-agency communication and collaboration between wildlife management agencies in different states (e.g. the IDFG and Colorado Division of Wildlife and Parks, Montana Fish Wildlife and Parks) was also important in building an appreciation for collaboration with large-scale datasets. In our case study in particular, the IDFG previously helped to synthesise mule deer demographic data across multiple states in the western USA (Unsworth *et al.* 1999). This helped pave the way for collaboration with universities, and vice versa.

There was also serendipity in our case study. In 2008, Mark Hebblewhite received a NASA ecological forecasting grant, focused on research. His project goal was to understand global-scale ungulate population dynamic responses to climate change, with a focus on developing integrated population models for North American elk, and caribou, across their circumpolar distribution (Post *et al.* 2009, Ahrestani *et al.* 2013, 2016). While this research provided interesting ecological insights, it was not initially related to management. The NASA-funded biodiversity research team, however, including the authors of this chapter, leveraged the funding to support related collaborative syntheses, which similarly demonstrated the benefits of remote sensing data in

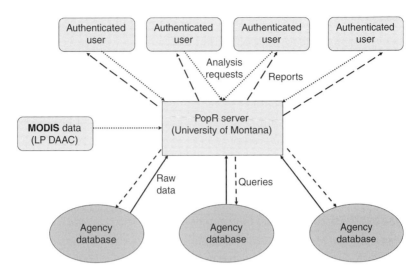

FIGURE 7.6 Graphical representation of the PopR distributed network architecture. Arrows represent the workflow from user to server to state wildlife agency database and back to the user. Remote sensing data improves the prediction of key demographic parameters (in this case study, overwinter survival of mule deer fawns) at the analysis stage, and MODIS data can be included directly from the NASA LP DAAC or other post-processed data sources, stored locally or on agency or user servers.

understanding large-scale regional ungulate ecology (Griffin *et al.* 2011, Brodie *et al.* 2013, Lukacs *et al.* 2018) through inter-agency collaborative workshops, facilitated in part by the Montana Cooperative Wildlife Research Unit. And through the process of working with NASA, the author Mark Hebblewhite transferred this knowledge and excitement to his graduate students and colleagues at the University of Montana. One such doctoral student, Mark Hurley, was also the mule deer research wildlife biologist for the IDFG. Mark's exposure to the innovative and creative thinking in the IDFG research culture encouraged the uptake of these breakthroughs in remote sensing science. This highlights the role that NASA's basic research funding has in kick-starting novel applications of remote sensing data to conservation and management.

Scientific publications themselves are insufficient to successfully translate research into management. A crucial missing step is the implementation of easy-to-use tools (software) to make statistical models available. Unfortunately, this step is not typically pursued by university academics. For our application, we developed customisable software directly with IDFG biologists and managers, with funding and support provided primarily by the IDFG. We then tested software interfaces in collaborative workshops, demonstrations, and with continued feedback between managers and researchers. Through this process we developed the PopR software (Nowak *et al.* 2017), using the underlying R programming language (R Core Team 2014). Modern computer software makes wildlife management and research easier and allows increasingly complex tasks to become routine, all of which lowers the barrier to initiating such customisable software applications. Unfortunately, data storage and reporting rarely keep pace with the rapid expansion of data-analysis software. Such disconnects in workflow can lead to missed opportunities, where data are not used to their fullest extent and new information is slow to emerge. Led by Paul Lukacs and Josh Nowak (University of Montana) in collaboration with the IDFG, PopR (https://popr.cfc.umt.edu, Figures 7.6 and 7.7) merges wildlife management agency databases with state-of-the-art statistical software for real-time wildlife data analysis, population modelling, and reporting (Figure 7.7).

The interface to PopR is a secure website, allowing access from any location and from any platform (personal computer, smartphone, tablet, etc.). PopR connects to remote data sources through an application program interface hosted at the University of Montana (Figure 7.6). It easily implements the kinds of Bayesian IPMs combining multiple data sources that we briefly described here (Equation 1, in Hurley 2016). A major advantage of this PopR interface is that it does not require specialised knowledge or training in Bayesian statistics or R programming (the programming language behind PopR). It is accessed through a drop-down graphic user interface that requires minimal training to successfully use. PopR also implements individual data-source analyses such as survival, detectability, herd composition, harvest, among other data sources. The interface can

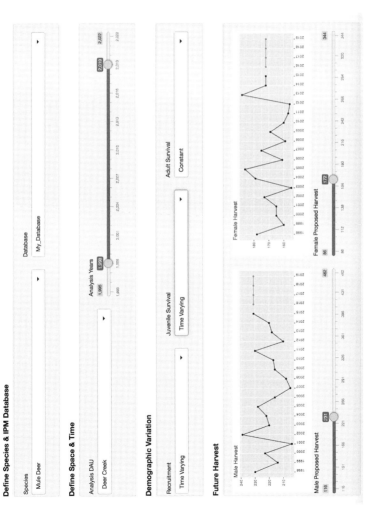

FIGURE 7.7 An example screenshot for mule deer population modelling in the Deer Creek Deer Analysis PMU in Idaho from the PopR software interface. Here, the analyst sees a graphic user interface that links to IDFG databases stored on a secure agency server (Figure 7.6) to conduct an analysis of changes in harvest under different assumptions of recruitment, mule deer survival rates, and different harvest levels. See https://popr.cfc.umt.edu/Demo/ for more examples.

seamlessly integrate with NASA's (and other) remote sensing data storage archives, such as the LP DAAC's archive for MODIS-type data. Finally, PopR generates reports and figures for rapid dissemination and incorporation into wildlife decision-making processes. It presents a seamless workflow from data to analysis to reporting. Other wildlife management agencies are starting to take up the PopR approach for wildlife management, and working with researchers at the University of Montana to develop customisable remote sensing applications, such as this case study described here. The website offers free demonstrations for interested readers (https://popr.cfc.umt.edu). We feel that the success of PopR in the IDFG and beyond was largely a result of the software development process, which was a collaborative one together with IDFG research biologists and managers, as described above. If researchers want software products to be used, we believe that collaborative development with the end-user is a crucial ingredient.

7.8   CONCLUSIONS

In summary, we have described a novel case study where we developed rigorous, regional-scale survival and population models for mule deer across Idaho. These survival and population models incorporated MODIS remote sensing data on vegetation structure (NDVI) and snowcover to successfully predict wildlife population dynamics and harvests in a manner that could be incorporated into real-time harvest management. Our models will provide a platform for the management of mule deer harvests in Idaho, and could even have predictive power in adjacent states such as Montana, Utah, eastern Washington, and Wyoming. More importantly, our approach provides a template for the management of other wildlife that are harvested in a similar way.

We identify the steps that we felt were important to achieve success in this case study. To reiterate, these were a productive wildlife agency culture regarding research and management, which includes two-way relationships between research and management, freedom to develop innovative approaches, a culture that adopts the highest scientific standards in management, and productive university–research collaborations. Second, it was critical to have a standardised database structure for wildlife management and wildlife inventory data, as well as standardised survey protocols, which often come from a centralising process. This led to the third step, rigorous population-based science to support population modelling and ecological forecasting. The resulting scientific publications, although a fourth important step, were insufficient to make the science useable by management or conservation agencies. The fifth step was thus agency collaboration, communication, analysis workshops, and demonstrations, to obtain guidance from wildlife managers about what tools would be useful. University research and culture is often not well suited to this level of engagement, but in our case study it was perhaps the most critical link to both 'sell' and receive feedback on how to make the science useful to wildlife managers, the ultimate end-user. Finally, building on the critical advice of this final step, we then developed customisable software, based on close development with wildlife managers (the end-user). Without this useable tool, code would have remained 'black box', limited only to scientists, and unable to be used and implemented in a management context.

ACKNOWLEDGEMENTS

We thank the editors of this volume, and an anonymous reviewer, for comments and reviews that improved the manuscript. Funding for this study was provided by the Idaho Department of Fish and Game (IDFG), Federal Aid in Wildlife Restoration Grant number W-160-R-37, NASA grant number NNX11AO47 G, University of Montana, Mule Deer Foundation, Safari Club International, Deer Hunters of Idaho, Universite Lyon, and Foundation Edmund Mach, Trentino, Italy. We thank dozens of IDFG biologists, supervisors, and wildlife technicians who collected the invaluable field data. We thank S. Running, J. St.Peter, W. K. Smith, and M. Zhao for remote sensing assistance.

REFERENCES

Ahrestani, F. S., Hebblewhite, M., and Post, E. S. (2013). The importance of observation versus process error in analyses of global ungulate populations. *Scientific Reports*, **3**, 03125.

Ahrestani, F. S., Hebblewhite, M., Smith, B., Running, S. W., and Post, E. (2016). Dynamic complexity and stability of herbivore populations at the species distribution scale. *Ecology*, **97**, 3184–3194.

Apollonio, M., Andersen, R., and Putman, R. (2010a). *European Ungulates and Their Management in the 21st Century*. Cambridge: Cambridge University Press.

Apollonio, M., Andersen, R., and Putman, R. (2010b). *Present Status and Future Challenges for European Ungulate Management*. Cambridge: Cambridge University Press.

Besbeas, P., Freeman, S. N., Morgan, B. J. T., and Catchpole, E. A. (2002). Integrating mark–recapture–recovery and census data to estimate animal abundance and demographic parameters. *Biometrics*, **58**, 540–547.

Bishop, C. J., White, G. C., Freddy, D. J., Watkins, B. E., and Stephenson, T. R. (2009). Effect of enhanced nutrition on mule deer population rate of change. *Wildlife Monographs*, **172**, 1–28.

Borowik, T., Pettorelli, N., Sonnichsen, L., and Jedrzejewska, B. (2013). Normalized difference vegetation index (NDVI) as a predictor of forage availability for ungulates in forest and field habitats. *European Journal of Wildlife Research*, **59**, 675–682.

Brashares, J. S., Arcese, P., Sam, M. K., et al. (2004). Bushmeat hunting, wildlife declines, and fish supply in West Africa. *Science*, **306**, 1180–1183.

Brodie, J., Johnson, H. E., Mitchell, M. S., et al. (2013). Relative influence of human harvest, carnivores and weather on adult female elk survival across western North America. *Journal of Applied Ecology*, **50**, 295–305.

Caswell, H. (2000). Prospective and retrospective perturbation analyses: their roles in conservation biology. *Ecology*, **81**, 619–627.

Clark, S. G. and Miloy, C. (2014). The North American model of wildlife conservation: an analysis of challenges and adaptive options. In Clark, S. G. and Rutherford, M.B., eds., *Large Carnivore Conservation: Integrating Science And Policy In The North American West*. Chicago, IL: University of Chicago Press, pp. 289–324.

Cote, S. D., Rooney, T. P., Tremblay, J. P., Dussault, C., and Waller, D. M. (2004). Ecological impacts of deer overabundance. *Annual Review of Ecology Evolution and Systematics*, **35**, 113–147.

Eldenshink, J. (2006). A 16-year time series of 1 km AVHRR satellite data of the conterminous United States and Alaska. *Photogrammetry Engineering and Remote Sensing*, **72**, 1027–1035.

Fryxell, F. M., Packer, C., McCann, K. S., Solberg, E. J., and Saether, B. E. (2010). Resource management cycles and the sustainability of harvested wildlife populations. *Science*, **328**, 903–907.

Gaillard, J.-M., Festa-Bianchet, M., Yoccoz, N. G., Loison, A., and Toigo, C. (2000). Temporal variation in fitness components and population dynamics of large herbivores. *Annual Review of Ecology and Systematics*, **31**, 367–393.

Gordon, I., Hester, A. J., and Festa-Bianchet, M. (2004). The management of wild large herbivores to meet economic, conservation and environmental objectives. *Journal of Applied Ecology*, **41**, 1021–1031.

Griffin, K., Hebblewhite, M., Zager, P., *et al.* (2011). Neonatal mortality of elk driven by climate, predator phenology and predator diversity. *Journal of Animal Ecology*, **80**, 1246–1257.

Hall, D. K., Riggs, G. A., Salomonson, V. V., DiGirolamo, N. E., and Bayr, K. J. (2002). MODIS snow-cover products. *Remote Sensing of Environment*, **83**, 181–194.

Hebblewhite, M., Merrill, E. H., and McDermid, G. (2008). A multi-scale test of the forage maturation hypothesis for a partially migratory Montane elk population. *Ecological Monographs*, **78**, 141–166.

Hobbs, N. T. (1996). Modification of ecosystems by ungulates. *Journal of Wildlife Management*, **60**, 695–713.

Hobbs, N. T. and Hooten, M. B. (2015). *Bayesian Models: A Statistical Primer for Ecologists*. Princeton, NJ: Princeton University Press.

Huete, A., Didan, K., Miura, T., *et al.* (2002). Overview of the radiometric and biophysical performance of the MODIS Vegetation indices. *Remote Sensing of Environment*, **83**, 195–213.

Hurley, M. A., ed. (1992). *Aerial Population Surveys. Blackfoot–Clearwater Elk Study – Progress Report*. Helena, MT: Montana Department of Fish, Wildlife, and Parks.

Hurley, M. A. (2016). *Mule Deer Population Dynamics in Space and Time: Ecological Modeling Tools for Managing Ungulates*. Missoula, MT: University of Montana.

Hurley, M. A., Unsworth, J. W., Zager, P., Hebblewhite, M., *et al.* (2011). Demographic response of mule deer to experimental reduction of coyotes and mountain lions in southeastern Idaho. *Wildlife Monographs*, **178**, 1–33.

Hurley, M. A., Hebblewhite, M., Gaillard, J. M., *et al.* (2014). Functional analysis of normalized difference vegetation index curves reveals overwinter mule deer survival is driven by both spring and autumn phenology. *Philosophical Transactions of the Royal Society of London B*, **369**, 20130196.

Hurley, M. A., Hebblewhite, M., Lukacs, P. M., *et al.* (2017). Regional-scale models for predicting overwinter survival of juvenile ungulates. *The Journal of Wildlife Management*, doi: 10.1002/jwmg.21211.

Idaho Department of Fish and Game (2013). *Big Game Harvest Statewide, Idaho Department of Fish and Game*. Boise, ID: Idaho Department of Fish and Game.

Johnson, H. E., Mills, L. S., Stephenson, T. R., and Wehausen, J. D. (2010). Population-specific vital rate contributions influence management of an endangered ungulate. *Ecological Applications*, **20**, 1753–1765.

Kaplan, E. L. and Meier, D. B. (1958). Nonparametric estimation from incomplete observations. *Journal of the American Statistical Association*, **53**, 457–481.

Kery, M. and Schaub, M. (2012). *Bayesian Population Analysis Using WinBUGS: A Hierarchical Perspective*. San Diego, CA: Acadmic Press.

Leopold, A., ed. (1933). *Game Management*. New York, NY: Charles Scribner's Sons.

Lukacs, P. M., Mitchell, M. S., Hebblewhite, M. *et al.* (2018). Factors influencing elk recruitment across ecotypes in the Western United States. *Journal of Wildlife Management*, doi: 10.1002/jwmg.21438.

Monteith, K. L., Bleich, V. C., Stephenson, T. R., *et al.* (2014). Life-history characteristics of mule deer: effects of nutrition in a variable environment. *Wildlife Monographs*, **186**, 1–62.

Nowak, J. J., Lukacs, P. M., Hurley, M. A., *et al.* (2017). Customized software to streamline routine analyses for wildlife management. *Wildlife Society Bulletin*, doi: 10.1002/wsb.841.

Pettorelli, N. (2013). *The Normalized Difference Vegetation Index*, Oxford: Oxford University Press.

Post, E. S., Brodie, J., Hebblewhite, M., *et al.* (2009). Global population dynamics and hot spots of response to climate change. *Bioscience*, **59**, 489–499.

Quality Deer Management Association (2017). QDMA's whitetail report 2017: an annual report on the status of white-tailed deer. The Foundation of the Hunting Industry in North America. See www.qdma.com/wp-content/uploads/2017/03/WR-2017.pdf. Accessed 15 August 2017.

R Core Team (2014). *R: A Language and Environment for Statistical Computing*. Vienna: R Foundation for Statistical Computing.

Ramsay, R. and Silverman, B. W. (2005). *Functional Data Analysis*. New York, NY: Springer.

Ripple, W. J., Newsome, T. M., Wolf, C., *et al.* (2015). Collapse of the world's largest herbivores. *Science Advances*, **1**, e14000103.

Royle, J. A. and Dorazio, R. M. (2006). Hierarchical models of animal abundance and occurrence. *Journal of Agricultural Biological and Environmental Statistics*, **11**, 249–263.

Samuel, M. D., Garton, E. O., Schlegel, M. W., and Carson, R. G. (1987). Visibility bias during aerial surveys of elk in northcentral Idaho. *Journal of Wildlife Management*, **51**, 622–630.

Samuel, M. D., Steinhorst, R. K., Garton, E. O., and Unsworth, J. W. (1992). Estimation of wildlife population ratios incorporating survey design and visibility bias. *Journal of Wildlife Management*, **54**, 718–725.

Schaub, M. and Kery, M. (2012). Combining information in hierarchical models improves inferences in population ecology and demographic population analyses. *Animal Conservation*, **15**, 125–126.

Shallow, J. R. T., Hurley, M. A., Monteith, K. L., and Bowyer, R. T. (2015). Cascading effects of habitat on maternal condition and life-history characteristics of neonatal mule deer. *Journal of Mammalogy*, **96**, 194–205.

Silvy, N. J. (2012). *The Wildlife Techniques Manual: Volume 2: Management*, 7th edn. Baltimore, MD: John Hopkins Press.

Sinclair, A. R. E., Fryxell, J., and Caughley, G., eds. (2005). *Wildlife Ecology and Management*. Oxford: Blackwell Science.

United States Fish and Wildlife Service (2016). Service distributes $1.1 billion to state wildlife agencies to support conservation, outdoor recreation, and job creation. Press Release, March 7. See www.fws.gov/news/ShowNews.cfm?re f=service-distributes-$1.1-billion-to-state-wildlife-agencies-to-support-&_I D=35495. Accessed 15 August 2017.

Unsworth, J. A., Leban, F. A., Leptich, D. J., Garton, E. O., and Zager, P., eds. (1994). *Aerial Survey: User's Manual*, 2nd edn. Biose, ID: Idaho Department of Fish and Game.

Unsworth, J. A., Pac, D. F., White, G. C., and Bartmann, R. M. (1999). Mule deer survival in Colorado, Idaho, and Montana. *Journal of Wildlife Management*, **63**, 315–326.

Unsworth, J. W., Kuck, L., and Garton, E. O. (1990). Elk sightability model validation at the National Bison Range, Montana. *Wildlife Society Bulletin*, **18**, 113–115.

White, M. A., de Beurs, K. M., Didan, K., *et al.* (2009). Intercomparison, interpretation, and assessment of spring phenology in North America estimated from remote sensing for 1982–2006. *Global Change Biology*, **15**, 2335–2359.

Williams, B. K., Nichols, J. D., and Conroy, M. J., eds. (2002). *Analysis and Management of Animal Populations*. New York, NY: Academic Press.

Zhang, X. Y., Friedl, M. A., Schaaf, C. B., *et al.* (2003). Monitoring vegetation phenology using MODIS. *Remote Sensing of Environment*, **84**, 471–475.

8 Lessons Learned from WhaleWatch

A Tool Using Satellite Data to Provide Near-Real-Time Predictions of Whale Occurrence

Helen Bailey, Elliott Hazen, Bruce Mate, Steven J. Bograd, Ladd Irvine, Daniel M. Palacios, Karin A. Forney, Evan Howell, Aimee Hoover, Lynn DeWitt, Jessica Wingfield, and Monica DeAngelis

8.1 INTRODUCTION

Despite a global ban on commercial whaling since 1986, the greatest threat to whales is still from human activity. Many whale populations have been slow to increase in number, and one reason for this is continued human-induced mortality (Thomas *et al.* 2016). Whales may be injured or killed when they collide with ships (Laist *et al.* 2001, Van Waerebeek *et al.* 2007), become entangled in fishing gear (Read *et al.* 2006, Moore *et al.* 2009), or are exposed to loud underwater noise (Nowacek *et al.* 2007, Di Iorio and Clark 2010). Along the California coast alone, there were 21 strandings of dead blue whales (*Balaenoptera musculus*) from 1988 to 2007, eight of which were confirmed as ship strikes (Berman-Kowalewski *et al.* 2010). In the decade between January 2007 and May 2017, there were 12 reported blue whale strandings off California (from the Marine Mammal Health and Response Program National Database). Ten of these strandings were confirmed as ship-strike mortalities (Figure 8.1). Often, though, the body of a dead blue whale will sink at sea and remains undetected, and unless the vessel reports that it struck the animal, it is not included in stranding records. In addition, confirming cause of death as a ship strike is often difficult because it requires a necropsy (an examination of

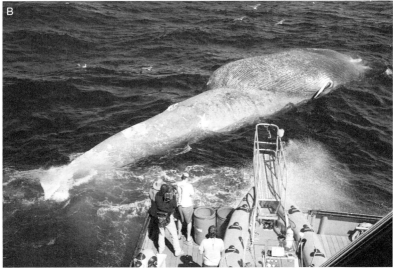

FIGURE 8.1 (A) A blue whale surfacing near the shipping channel off southern California. (A black and white version of this figure will appear in some formats. For the colour version, please refer to the plate section.) (B) A dead blue whale being examined by Oregon State University (OSU) researchers after being killed by a ship strike. (Photo by Craig Hayslip, OSU Marine Mammal Institute).

the whale's body), which is challenging to conduct on a large whale even if the whale washes up onshore. Therefore, the actual number of whales that are struck by ships is likely to be greatly underestimated (Laist *et al.* 2001, Kraus *et al.* 2005).

Entanglement in gear is another threat to large whales, particularly humpback whales (*Megaptera novaeangliae*) off the US west coast. Pot and trap fisheries were the most commonly recorded source of serious injury and mortality of humpback whales in US west coast waters, with 16 entanglements between 2007 and 2011 (Carretta *et al.* 2013). Such reports have been increasing in recent years, with 35 reported entanglements of humpback whales in 2015 and a record high of 54 in 2016 (National Marine Fisheries Service 2017). In 2015, there was a confirmed case of a blue whale that was entangled in fishing gear off California and, in 2016, there were four reports of blue whale entanglements (National Marine Fisheries Service 2017). There had previously been no reports of entangled blue whales in US west coast waters (Carretta *et al.* 2016). It was considered possible that blue whales were sufficiently large to swim and break through fishing gear without getting entangled in nets and rope lines (Carretta *et al.* 2016). Unfortunately, even the largest animal on the planet is not immune to this threat.

Blue whales are currently listed as Endangered on the International Union for Conservation of Nature's (IUCN) Red List and under the US Endangered Species Act. They can reach more than 30-m in length and weigh up to 180 tonnes. Blue whales have a specialist diet, feeding almost entirely on krill. They were heavily hunted following technological advances in commercial whaling operations in the late nineteenth century (Clapham *et al.* 1999). Globally, the total number of blue whales is roughly estimated at 10,000–25,000, although there is considerable uncertainty in various regions (Reilly 2008). The best-studied eastern North Pacific population is currently estimated at 1,647 blue whales, and has shown no signs of increase since surveys began in the 1990s (Calambokidis and Barlow 2004, 2013).

Over the last several decades, some marine mammal populations have been increasing off the US west coast as a result of federal and state protections. For some large whales, including the humpback and grey (*Eschrichtius robustus*) whale, population recovery has led to a re-evaluation of their protected status under the US Endangered

Species Act, whereas others, including the blue whale, are considered globally endangered throughout their range. Given the continued human-induced mortality and injury of whales in US west coast waters, the Protected Resources Division of the National Oceanic and Atmospheric Administration (NOAA), National Marine Fisheries Service (NMFS) West Coast regional office made it a priority to reduce the number of ship strikes and entanglements in fishing gear for large whales. The US Department of Commerce, through NOAA/NMFS, is charged with protecting cetaceans (whales, dolphins, and porpoises), seals, and sea lions, and implementing the Marine Mammal Protection Act, Endangered Species Act, and other federal regulations. NOAA works with interested stakeholders to conserve and protect marine mammals and their habitats through science-based decision-making and compliance with regulations. Each year, the distribution of whales off the west coast can shift, based on a variety of biological and environmental conditions, and NOAA considers all of the available information, along with data collected on entangled and ship-struck animals, in its effort to reduce the number of entanglements and vessel collisions. Blue whales were considered a high priority species for the NOAA/NMFS West Coast regional office because the eastern North Pacific population did not appear to be increasing (Calambokidis and Barlow 2013). The potential biological removal (PBR) estimated by the NOAA for this population, which is the maximum number of animals not including natural mortalities that can be sustainably removed from the population, is 2.3 blue whales per year (Carretta *et al.* 2016).

There is a need for finer-scale information on the distribution and density of marine mammals. To meet this need, the NOAA/ NMFS Southwest Fisheries Science Center (SWFSC) had previously developed habitat-based density models for blue whales and other cetaceans (including other large baleen whales, toothed whales, and dolphin species), based on shipboard surveys conducted along the US west coast from 1991 to 2008 (Becker *et al.* 2012a, 2012b, Forney *et al.* 2012, Redfern *et al.* 2013, Becker *et al.* 2016). These spatially explicit models predicted the expected number of animals across a surface of grid cells, based on associations with ocean conditions (Redfern *et al.* 2006). These models had been extensively validated and updated, and have been used in multiple management contexts (e.g. US Navy environmental assessments, assessing ship-strike risk to large whales), demonstrating the value of such habitat-based spatial-density models for the assessment and management of anthropogenic impacts to whales. However, these models only provided information for the summer and autumn period, when the surveys were conducted, and were unable to describe or account for the long-distance seasonal migrations of large whales (Stone *et al.* 1990) and year-round patterns of whale distribution and density. During the period from January 2007 to May 2017, nine blue whale strandings occurred in the autumn (September to November) and the remaining three strandings occurred in May, July, and August.

In response to these known data gaps, we believed our previous work, analysing the movements of Argos satellite-tracked blue whales (Bailey *et al.* 2009), could be further developed to assist in identifying year-round patterns of blue whale distribution and densities. The aim of the WhaleWatch project was to combine whale satellite telemetry data with satellite-derived environmental data to predict whale occurrence and densities, based on the latest environmental conditions (Figure 8.2). In response to a grant solicitation by the National Aeronautics and Space Administration (NASA), we built a multi-institutional team of academic groups and governmental organisations. The NASA Applied Sciences Program, part of the Earth Science Division, funded the project in 2011.

In the development of the project plan, we recognised that a key advantage of satellite telemetry data over visual surveys is that the tracks provide an animal's-eye view of critically important areas and include year-round information on movement patterns. An understanding of how the environment influences animal movements and distribution allows predictions to be made about their occurrence relative to the ocean conditions. This provides a scientific basis to answer the question that had been raised by the shipping and fishing

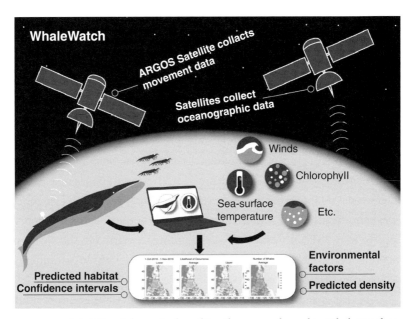

FIGURE 8.2 Schematic describing the approach used in WhaleWatch to combine Argos satellite telemetry data for whales with satellite-derived environmental data, to determine the factors influencing blue whale distribution and to predict their occurrence and densities in a near-real-time tool.

industries, 'Where are the whales most likely to be at any given time?', which the NOAA/NMFS West Coast regional office had been unable to adequately address.

As far as we were aware, tools had been implemented to create an online mapping portal of marine mammal densities, based on models of sightings data (Best *et al.* 2012), but these maps gave seasonal averages and were not updated according to current ocean conditions. There are systems that detect whales in real time via passive acoustic monitoring (Van Parijs *et al.* 2009, Baumgartner *et al.* 2013) or from sightings from aerial or boat-based surveys (Brown *et al.* 2007), but a near-real-time tool for predicting whale occurrence and density was not available. A tool had previously been developed to reduce bycatch interactions between loggerhead turtles (*Caretta caretta*) and pelagic longline fisheries in the central Pacific (Howell *et al.*

2008). This tool, known as TurtleWatch (https://www.pifsc.noaa.gov/eod/turtlewatch.php), combined satellite telemetry data for loggerhead turtles, bycatch information, and satellite-derived environmental data to identify high-risk areas. The tool indicates areas on a map that the Hawaii-based pelagic longline fishery can avoid to reduce the risk of interactions with loggerhead turtles, based on satellite-derived SST data for the most recently available 3-day period. A similar tool, based on an environment-driven model (termed habitat-preference model), had also been used to forecast the distribution of southern bluefin tuna (*Thunnus maccoyii*) in eastern Australian waters and forms the basis for their dynamic spatial management (Hobday *et al.* 2011). These dynamic management tools served as both the inspiration and a template for the development of our project, WhaleWatch.

8.2 WHALEWATCH PROJECT

The goal of WhaleWatch was to create an online, near-real-time tool, predicting whale occurrence and densities in US west coast waters, a known feeding area. The strongest driver of whale movements on the foraging grounds is likely to be their prey. Blue whales are specialist feeders on krill (pelagic marine crustaceans of the order Euphausiacea) (Sears and Perrin 2009) and the eastern North Pacific population forages off Mexico, the US west coast, and farther offshore (Mate *et al.* 1999, Calambokidis *et al.* 2008, Bailey *et al.* 2009). However, information on krill distribution and abundance is scarce, particularly concomitant to whale tracking data, and data are not collected with enough spatial range or temporal resolution to be used in near real time. Relationships have been found between krill distributions and oceanographic and bathymetric features along the California coast. Krill hotspots off California were in a northwest–southeast direction, occurring along the 200–2,000-m isobaths, which corresponded with the shelf break and slope habitat (Santora *et al.* 2011). Krill appeared to avoid locations of strong upwelling and aggregated in areas downstream of these locations (Santora *et al.* 2011), which are distinguished by reduced SST and increased phytoplankton, identified by chlorophyll-*a* concentrations. Therefore, environmental data that could be obtained from satellite remote sensing could be used as a proxy for prey distribution and to identify environmental factors that are directly or indirectly influencing blue whale occurrence.

In the WhaleWatch project, we analysed the blue whale satellite telemetry data to determine their distribution off the US west coast, identified predictable hotspots, characterised their migration patterns, and quantified seasonal and interannual variability in occurrence. Blue whale telemetry data were obtained from Bruce Mate's team at Oregon State University, who had tagged 171 blue whales, mainly off California, from 1993 to 2008 (Bailey *et al.* 2009, Irvine *et al.* 2014, Hazen *et al.* 2016). The Argos satellite positions obtained from the tagged whales were filtered and regularised at daily intervals using a state-space movement model. A state-space model predicts the future state (in this case the next location), based on the previous state (or previous location) (Patterson *et al.* 2008). The state-space movement model includes two equations that take into account the error distribution in the recorded Argos locations and a process model that describes how the movement changes over time (Patterson *et al.* 2008, Jonsen *et al.* 2013). The process model can include behavioural modes with different properties. For example, migration behaviour is typically characterised by faster speeds and a persistent direction whereas foraging behaviour within a prey patch may have slower speeds and higher turning angles. A matrix of switching probabilities to indicate the likelihood of changing from one behaviour to another is included in the process model and is termed a switching state-space model. Such a model was applied to the blue whale Argos satellite telemetry data (Jonsen *et al.* 2005, Bailey *et al.* 2008). The model output gave estimates of daily positions, their uncertainty (as the 95% credible limits), and behavioural mode (typical of migration/transiting or more localised movements, suggesting foraging or breeding) (Bailey *et al.* 2009).

For each individual track, kernel home ranges and core areas were generated and then compared amongst whales to identify where

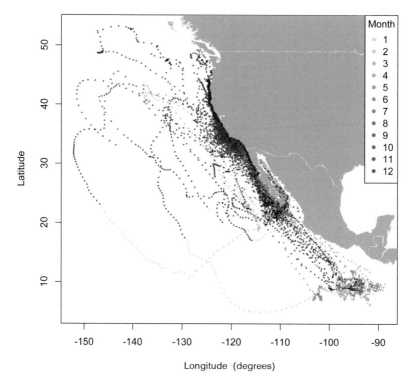

FIGURE 8.3 Daily state-space-model-derived positions for satellite-tagged blue whales, shaded by month.
Adapted from Hazen *et al.* 2016.

the highest overlap occurred to delineate high-use areas (Irvine *et al.* 2014). The tagged blue whales occurred throughout US west coast waters during all months of the year, but mainly in July to December (Figure 8.3). The greatest overlap near the shipping lanes, areas of increased concern for whale ship strikes from high ship traffic, occurred from July to October off southern California and during August to November off Central California (Irvine *et al.* 2014) (Figure 8.4).

The blue whale positions were generally concentrated along regions of high productivity along the continental shelf and upper slope, with whales exhibiting a strong seasonal pattern, migrating from high latitudes in the summer and autumn to lower latitudes in the winter and spring (Bailey *et al.* 2009, Irvine *et al.* 2014). We

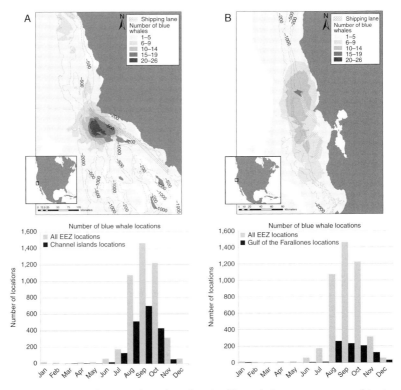

FIGURE 8.4 Number of overlapping blue whale core areas near shipping lanes off southern California (A) and central California (B). The inset maps show the location off the US west coast. Graphs show the number of blue whale locations in US west coast waters (grey) and in the area shown in the maps (black).
Adapted from Irvine *et al.* 2014. (A black and white version of this figure will appear in some formats. For the colour version, please refer to the plate section.)

therefore integrated the blue whale positions with environmental variables that were metrics of these processes and available for the duration of the tracking period and in near real time.

Since tracking data are presence-only information, meaning that they only tell you where the whales were and not where they could have gone, we needed a way to compare these data with the available habitat. We used a correlated random-walk model to simulate 'null' whales, using the duration, distance, and turning angles from the actual whale tracks. The whale positions (case points) and simulated positions (control points) were integrated with remotely sensed environmental data and then used to develop a habitat-preference model.

The satellite-derived environmental variables included dynamic oceanographic variables, such as SST, chlorophyll-a concentration, sea surface height (SSH) anomaly, wind speed, eddy kinetic energy (EKE), and Ekman upwelling (WEKM), as well as bathymetric variables, including water depth, seabed slope, and slope direction. The sources, resolutions, and time periods of these satellite-derived environmental products are described in Table 8.1. These variables were chosen because they provide key ecological links to blue whales, either directly through water column preferences or through the preferences of their prey, krill.

The SST can indicate water properties that blue whales use to cue their movements and migration. In addition, the SST may describe the physiological preferences of krill in the California Current (e.g. Croll *et al.* 2005). The standard deviation of the SST spatially gives a metric of gradient strength, or simply the mesoscale variability, including frontal structures and eddy edges. Chlorophyll-a is the primary pigment used for photosynthesis in phytoplankton and gives a measure of the concentration of phytoplankton (Horning *et al.* 2010). Areas with high chlorophyll-a concentration, where the colour of the water is greener, can be used to indicate areas of higher primary productivity and this has been shown as an important precursor to krill distributions in the California Current (Fiedler *et al.* 1998, Croll *et al.* 2005, Santora *et al.* 2012).

The SSH anomaly, the difference between the estimated SSH and mean sea surface, gives a metric of convergence and divergence in the ocean, when it is positive or negative, respectively. Eddy kinetic energy is the energy associated with turbulent parts of the ocean. It is calculated from the zonal (east–west) and meridional (north–south) geostrophic current components, which is the ocean flow arising from the pressure gradient force that is balanced by the

Table 8.1 *Summary of satellite-derived remotely sensed environmental data used for WhaleWatch. Adapted from Hazen et al. 2016.*

Variable	Product/Sensor	Grid resolution	Temporal coverage	Source	Weblink (citation)	Mechanism
Sea surface temperature and standard deviation (SST, SSTsd)	AVHRR Pathfinder v. 5 (day and night)	4.4 km 5.5-km	15 September 1994–13 April 2008	NOAA/ NESDIS	https://pathfinder.nodc.noaa.gov (Casey et al. 2010)	SST: physiological driver of prey distribution
	Reynolds Optimum Interpolation v.2 (AMSR, AVHRR, *in situ*)	25-km (0.25-degree)	1 September1981– present	NOAA/ NCDC	http://rda.ucar.edu/data sets/ds277.7/docs/daily-ss t.pdf (Reynolds et al., 2007)	SSTsd: indicator of mesoscale activity, e.g. frontal variability
	MODIS/Aqua	0.0125-degree on ERDDAP	5 July 2002– present	NASA/ GSFC	https://coastwatch.pfeg.no aa.gov/infog/MW_sstd_las .html (Aqua MODIS information at https://oceancolor.gsfc.n asa.gov)	
Chlorophyll-*a* concentration (Chl-a)	SeaWiFS/ Orbview-2	8.8-km	9 August 1998–10 April 2008	NASA/ GSFC	https://coastwatch.pfeg.no aa.gov/infog/SW_chla_las .html (O'Reilly et al. 1998)	Chl-a: Indicator of phytoplankton concentration

Table 8.1 (*cont.*)

Variable	Product/Sensor	Grid resolution	Temporal coverage	Source	Weblink (citation)	Mechanism
	MODIS/Aqua	4.4-km	5 July 2002–present	NASA/GSFC	https://coastwatch.pfeg.noaa.gov/infog/MW_chla_las.html (Aqua MODIS information at https://oceancolor.gsfc.nasa.gov)	and primary productivity
Sea surface height, standard deviation, and Eddy kinetic energy (SSH, SSHsd, EKE)	Merged (Topex/Poseidon, ERS-1/-2, Geosat, GFO, Envisat, Jason-1/-2) AVISO	0.25-degree	14 October 1992–present	AVISO/CMEMS	Formerly obtained from www.aviso.oceanobs.com. The Ssalto/Duacs altimeter products are now produced and distributed by the Copernicus Marine and Environment Monitoring Service (CMEMS) (http://marine.copernicus.eu/)	SSH: Indicator of aggregation SSHsd, EKE: Indicator of mesoscale variability
Wind velocity and Ekman upwelling (u-wind, WEKM)	SeaWinds/QuikSCAT	12.5-km	25 July 1999–18 November 2009	NASA/JPL	https://coastwatch.pfeg.noaa.gov/infog/QS_taux_las.html	uWind and WEKM: Indicator of upwelling strength

Table 8.1 (*cont.*)

Variable	Product/Sensor	Grid resolution	Temporal coverage	Source	Weblink (citation)	Mechanism
	ASCAT	25-km (0.25-degree)	16 October 2009–16 June 2013 (end date for monthly composite)	Eumetsat and NOAA/NESDIS	(NASA/JPL Winds information at http://winds.jpl.nasa.gov) https://coastwatch.pfeg.noaa.gov/infog/QA_ux10_las.html	
				Eumetsat and NOAA/NESDIS	(Information by NOAA/NESDIS at http://manati.orbit.nesdis.noaa.gov/products/ASCAT.php)	
Bottom depth, rugosity, aspect, and slope (bathy, bathysd, aspect, slope)	SRTM30_PLUS v.6.0 digital bathymetry	0.0083-degree	Fixed	UCSD/SIO	http://topex.ucsd.edu/WWW_html/srtm30_plus.html (Becker *et al.* 2009)	Bathy: water column preferences. bathysd, aspect, slope:
Distance to shelf break (200-m isobath)	ETOPO2 v.2g	0.0333-degree	Fixed	NOAA/NGDC	http://www.ngdc.noaa.gov/mgg/global/relief/ETOPO2/ETOPO2v2-2006/ETOPO2v2g (National Geophysical Data Center 2006)	bathymetric influences on upwelling and prey distribution

Coriolis effect. The standard deviations of SSH and EKE are derived products that also describe mesoscale features, such as eddies, that may serve as aggregators of primary productivity or prey, directly. The WEKM is a derived product from wind stress that gives an estimate of vertical velocity, or the strength of wind-driven upwelling, which drives primary productivity (Croll *et al.* 2005). Bathymetric variables are particularly important in driving krill distributions (Santora *et al.* 2011), and the 200-m isobath has been shown to be related to blue whale habitat use (Redfern *et al.* 2013, Irvine *et al.* 2014). This shelf break serves to drive upwelling of nutrients into the photic zone, but also serves as a refuge for krill at depth.

A telemetry-based habitat model was developed by fitting a generalised additive mixed model (GAMM) to the blue whale positions (case points) and simulated positions (control points) in a case–control design, using the environmental data as predictor variables (Hazen *et al.* 2016). The best-fit models included the environmental variables SST, chlorophyll-*a* concentration, SSH anomaly standard deviation, water depth, and the standard deviation of water depth (indicating the seabed slope). Multiple models were created to determine if the relationships were robust to the selection of the control points and to estimate the mean and standard deviation of the prediction values for the ensemble model runs (Hazen *et al.* 2016). The results of 40 models that were run with different control points showed that the same environmental variables were consistently identified as statistically significant. This indicated that the models were robust to the selection of the control points.

Once the final habitat models had been developed, the process was automated to download the latest environmental data using NOAA CoastWatch–West Coast node's ERDDAP data server (Simons 2016) and AVISO/CMEMS (Copernicus Marine and Environment Monitoring Service) for the altimetry data. The ERDDAP data server (https://coastwatch.pfeg.noaa.gov/erddap) gives a simple, consistent way to download and grid subsets of oceanographic data. The ESA CMEMS (http://marine.copernicus.eu) programme provides full and open access to data and information related to the state of the physical oceans and regional seas. Once the data were downloaded, they were then re-gridded and input into the model to create the monthly prediction estimates. The monthly predictive maps are served, via restful URL, automatically to the NOAA/NMFS West Coast Region website (www.westcoast.fisheries.noaa.gov/whalewatch).

Serving the WhaleWatch product on a publicly accessible NOAA website provides information to managers, the shipping industry, fishers, and other stakeholders in near real time (Figure 8.5). The current predictive maps are displayed on the home page and, when a new image is created with the latest environmental data, older images are moved to the archive page to remain available. The interannual variability in blue whale use of US west coast waters (Barlow and Forney 2007, Forney *et al.* 2012) highlights the benefit of having a near-real-time tool that can account for changes in ocean conditions (Hazen *et al.* 2016) (Figure 8.6). Coinciding with our model predictions, anecdotal evidence indicates that there were fewer blue whales than average in 2015, with whale-watching companies reporting half the number of blue whales in 2015, a year with anomalously warm water temperatures, than they had in previous years (M. DeAngelis, personal communication, 2016).

The maps, associated environmental data, and estimated prediction values can be viewed and downloaded. They provide a decision-support tool to help managers determine if action is necessary to mitigate the risk of human impacts in blue whale high-use areas. Measuring the actual effect of the WhaleWatch tool on reducing ship strikes of blue whales is difficult given the small number of collisions reported annually. However, the WhaleWatch tool aims to provide information on expected high-use and high-density hotspots to help inform decision-making on whether mitigation measures, such as vessel speed restrictions, should be considered. The probability of a lethal injury during a ship strike is related to vessel speed (Vanderlaan

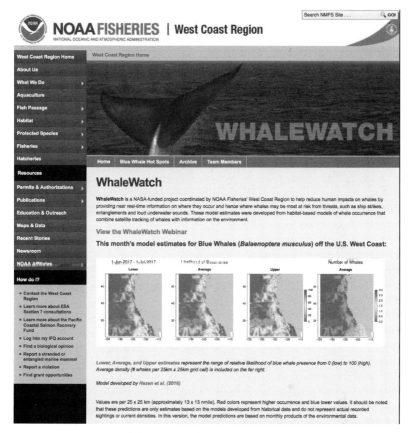

FIGURE 8.5 WhaleWatch webpage hosted on the NOAA/NMFS West Coast Region website (www.westcoast.fisheries.noaa.gov/whalewatch, accessed 6 July 2017). (A black and white version of this figure will appear in some formats. For the colour version, please refer to the plate section.)

and Taggart 2007, Wiley *et al.* 2011), and mandatory speed limits of 10 knots in Seasonal Management Areas have been implemented along the US east coast (Laist *et al.* 2014). The model developed in WhaleWatch used blue whale data, which were a high priority species for NOAA, but now that it has been developed, the framework could be applied to other species with tracking data in the future, such as the humpback whale, which has been found to have high rates of entanglements with fishing gear.

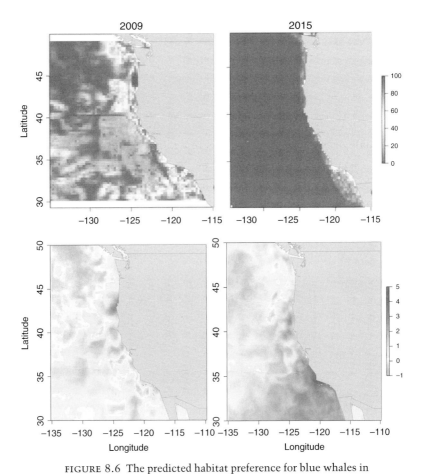

FIGURE 8.6 The predicted habitat preference for blue whales in September 2009 (an average year for ocean conditions) and September 2015 (an unusually warm period) off the US west coast (upper panels, percent likelihood of occurrence) and the corresponding satellite-derived SST anomaly for those periods (lower panels, °C).

Adapted from Hazen *et al.* 2016. (A black and white version of this figure will appear in some formats. For the colour version, please refer to the plate section.)

8.3 LESSONS LEARNED

The availability of near-real-time data, such as satellite-derived environmental measurements, has led to the growth of the concept of dynamic ocean management (Hobday *et al.* 2014, Maxwell *et al.*

2015), which 'uses near real-time data to guide the spatial distribution of commercial activities' (Lewison *et al.* 2015). TurtleWatch has been a pioneer for the concept of dynamic ocean management (Howell *et al.* 2008, 2015). TurtleWatch displays a map of near-real-time SST, whereas a key innovation of WhaleWatch is that it involves processing multiple environmental variables in a habitat-preference model to predict whale occurrence and densities in a monthly product. This more complex approach led to additional challenges and complications that we had not initially anticipated, but the approach was necessary to adequately describe habitat use and migratory behaviour. This approach is also likely to be required for many other marine species for which a single environmental proxy, such as SST, may not be sufficient to describe and predict their distribution.

As the number of projects creating a near-real-time tool to assist in natural-resource management is increasing, we describe these six lessons learned, and critical components, in more detail below to assist such projects in the future. Key lessons that we learned and critical components during this process were:

(i) Involvement of end-users in product development: the importance of incorporating user input from the start of the project, and during its development, rather than just delivering the final product at the end.

(ii) Big data by small increments: sufficiently large sample sizes may take many years to collect, but are necessary to characterise species distributions.

(iii) Data availability for the past, present and future: environmental data form a critical component of these models and tools, requiring long-term continuity in these remotely sensed environmental variables.

(iv) Biological realism versus statistical model fit: although assessing the model fit is an important part of evaluating the statistical habitat-preference model, it is critical that the predictions are biologically realistic.

(v) Validation: What is the truth?: all observation methods have limitations, but similarities and differences in the results from different approaches can provide further insight into the patterns and processes.

(vi) Automation and sustainability: sustainability of the near-real-time tool beyond the lifetime of the project generally requires automation so there is minimal human intervention and maintenance (and consequently funding) necessary.

(i)   Involvement of End-Users in Product Development

End-users are most likely to use products that meet their needs, that more easily allow a task to be completed or a decision to be made, and that are compatible with the existing data, tools, and software or platform that they are already using. In our project proposal for WhaleWatch, we discussed the idea with the end-user, the NOAA/NMFS West Coast regional office, from the beginning, and their support was critical for the success of the WhaleWatch project, including the transition from scientific research to management application. We developed a plan to use a three-phase approach to ensure that the products could easily be used by the NOAA/NMFS West Coast regional office, improving their ability to make policy and management decisions regarding the protection of large whales. This approach involved regular communication and a face-to-face meeting during the first year of the project, to identify the major information gaps and to determine how the WhaleWatch tool could most effectively help to fill them. We also discussed data formats and software program requirements to maximise the functionality and use of this tool by NOAA. Initially, we planned to develop two types of model: one modelling the presence of blue whales, to identify total habitat and to provide a measure of probability of occurrence (Edrén *et al.* 2010, Ainley *et al.* 2012), and the second using the behavioural mode from the switching state-space model to identify where animals were most likely to be foraging (similar to the approach in Bailey *et al.* 2012). The probability of occurrence estimates indicates areas where whales are potentially at high risk to anthropogenic threats and where whales are more vulnerable to ship strikes when they are feeding near the surface (Parks *et al.* 2012). However, during the first year of the project, the

NOAA Cetacean Density and Distribution Mapping Working Group (http://cetsound.noaa.gov/cda-index) established a data hierarchy. This ranked data types by their expected ability to accurately predict density or distribution, typically the most useful types of information to resource managers. This information hierarchy had five tiers, where tier 1 was considered the highest level and greatest value to resource managers and tier 5 the lowest.

In this hierarchy, telemetry data was only ranked as tier 4. Developing a model of probability of occurrence from the blue whale telemetry data would raise the information type to tier 3. Given the valuable movement and distribution information that is provided by telemetry data, as well as the time and financial investment, our aim was to process and format these data so that they would be raised to the highest level, tier 1. This required developing habitat-based density models. At the time, one of our team members was using telemetry data from harbour seals (*Phoca vitulina*) to estimate densities in proposed offshore wind-farm sites (Bailey *et al.* 2014). We decided to build on this approach (Aarts *et al.* 2008, 2012) to develop habitat-based density models from the blue whale telemetry data.

In the waters off the US west coast, habitat-based density models had been developed for cetaceans, based on sightings from line-transect surveys by the SWFSC (Forney *et al.* 2012). However, these surveys were only conducted from July to November, every 3–4 years, providing a single snapshot for those months of the year. Given the cost of surveys and limited at-sea days available, shipboard surveys could become even less frequent in the future. Our goal was to develop year-round estimates of habitat-based densities and regularly create new estimates, based on the latest environmental data. We would use the same spatial extent and grid cell size (25 × 25-km) as the existing SWFSC sightings-based models. The SWFSC had done a comparison of cetacean habitat models using *in situ* and remotely sensed environmental data. Becker *et al.* (2010) showed that satellite data (8-day temporal resolution and 5–35-km spatial resolution) had a predictive ability that was equal or greater than models developed with corresponding *in situ* data. The NOAA/NMFS West Coast regional office was also using remotely sensed SST for other management decisions, such as when considering the closure of the Pacific Loggerhead Conservation Area to the California drift gillnet fishery during a forecasted or occurring El Niño (50 CFR 660.713(c)(2)), when there are anomalously warm waters. The SWFSC and the NOAA/NMFS West Coast regional office therefore readily accepted and used remotely sensed environmental data.

As part of the modelling process, the SWFSC created maps of the standard error of the model predictions to give a measure of the error and uncertainty in the predictions of species distributions (Becker *et al.* 2010, Redfern *et al.* 2013). This had also been expressed as the confidence limits, which the SWFSC had presented along with the annual and multi-year average predictions to indicate the degree of interannual variability and model precision (Forney *et al.* 2012). As the products from WhaleWatch would be freely available on a public website to any stakeholders, we wanted to present our model uncertainty in a way that was easily interpretable by a general audience. We therefore decided we would present the average model prediction along with the lower and upper confidence intervals, to indicate the potential range of values that were plausible.

Having identified the requirements and objective of the WhaleWatch tool, during the second phase we developed and demonstrated the preliminary habitat model and its predictions at a meeting at the NOAA/NMFS West Coast regional office, midway through the project. Stakeholders, such as representatives from shipping companies and associations, the US Coast Guard, and the National Marine Sanctuaries also attended the presentation, which enabled us to obtain feedback from multiple potential end-users. As a result of this meeting, and to help facilitate the inclusion of stakeholder input and keep the team up to date on events, such as recent whale entanglements, the NOAA/NFMS West Coast regional office joined the monthly team calls for the remainder of the project.

In the third phase we incorporated this feedback into the final development and evaluation of the habitat model and automated the process into a near-real-time tool. This information was presented at the NOAA/NMFS West Coast regional office, which other stakeholders also attended. Having evaluated the final products, our team then worked closely with the NOAA/NMFS West Coast Region public affairs department and their website designer to create the WhaleWatch webpages on the NOAA/NMFS West Coast Region's website. This was followed by a public webinar, hosted by NOAA with a presentation by the team members, which described the project and the products and tool on the WhaleWatch webpages. The WhaleWatch webpages have had almost 1,000 views per year and a NASA ScienceCast video of the project had over 45,000 views in the first year.

The funding from NASA was essential for the development of WhaleWatch and the support of NOAA was critical for its incorporation into their decision-making process and for encouraging engagement by other stakeholders. This inter-institutional and inter-agency partnership brought together a range of data and skills to aid the NOAA/NMFS West Coast Region in achieving one of its goals of reducing human impacts on whales.

(ii) Big Data by Small Increments

It is difficult to observe the movements and behaviour of marine species, particularly those that occur farther offshore and undergo long-distance migrations. The use of satellite tracking for marine species has grown rapidly since the 1980s (Rutz and Hays 2009), but it is still financially costly and logistically challenging to tag large whales (Mate *et al.* 2007). The number of whales tagged each year depends on the funding available and the accessibility to whales in targeted areas. The large number of blue whale tracks analysed in the WhaleWatch project were obtained from tagging efforts that took place over a period of 15 years, funded by multiple sources.

We had initially hoped to create habitat models of humpback and grey whales too, but we found that the telemetry data available for these species was insufficient for the models, which required relatively large numbers of individuals and tracking days to adequately represent the population distribution.

There are many questions in movement ecology that telemetry data can help to answer (Hays *et al.* 2016). Tagging technology is rapidly developing to provide diving and acceleration data for large whales, in addition to horizontal movements. These data will provide information on their feeding behaviour, diving physiology, and response to anthropogenic activities and noises (Calambokidis *et al.* 2008, Goldbogen *et al.* 2013, Mate *et al.* 2016). However, it can take time to build up sufficient sample sizes and sample across multiple oceanographic regimes (e.g. Godley and Wilson 2008). Consequently, tracking large numbers of individuals in order to adequately capture population-level distribution patterns will generally require multi-year and multi-funding source projects. Collaboration and data sharing through networks, such as the Integrated Ocean Observing System Animal Telemetry Network (http://oceanview.pfeg.noaa.gov/ATN) and Ocean Biogeographic Information System (OBIS)–SEAMAP (http://seamap.env.duke.edu), will also be key to maximising sample sizes within and among species and habitats.

(iii) Data Availability for the Past, Present, and Future

Habitat models rely on an accurate sampling of the environmental characteristics that occur throughout the range of the organism being studied. As the size of an animal's range increases, the ability of traditional environmental sampling methods (i.e. surveys, fixed monitoring stations, etc.) to adequately describe its environmental characteristics decreases. Satellite-derived environmental data are often a critical component of modern habitat models for wide-ranging species as they provide global coverage that allows the entire range of an animal to be sampled. This is especially important for a very wide-ranging species such as the blue whale, which can travel over 100-km in a day (Mate *et al.* 1999). Further, the ocean is a highly variable environment over a wide range of spatial and temporal scales. The environmental characteristics of an area may vary over time from daily (e.g. tides), weekly (e.g. winds, storms), monthly (e.g. phytoplankton blooms), yearly (e.g. El Niño), and even decadally (e.g. Pacific Decadal Oscillation). The ability of satellite-derived environmental data to capture not only the environmental characteristics of an animal's range, but also how its range varies over time, is therefore crucial for the construction of comprehensive habitat-based species-distribution models.

The development of habitat-based models and near-real-time prediction tools requires that the environmental data are available not only for the observation period, but also for the current conditions, and into the future. The free and ready availability of satellite-derived environmental data has been a crucial component of the recent development in tools for dynamic ocean management (Hobday *et al.* 2014, Lewison *et al.* 2015, Maxwell *et al.* 2015). In order for such tools to continue to operate, it is necessary that there is continued long-term support for the satellites, sensors, data collection, storage, and processing that are required to provide such information. Consistent, science-quality data on SST, chlorophyll-*a* concentration, and SSH play a critical role in the success of these efforts. The development of data access protocols, such as the ERDDAP data server (Simons 2016), which provide a simple, consistent way to download subsets of scientific datasets in common file formats while also producing near-real-time graphs and maps, has facilitated the use of satellite-derived environmental data for novices and experts.

The loss of data services provided by NOAA would reduce the ability of such data to be assimilated and used by the wider community of scientists and managers. Near-real-time tools based on species' habitat preferences require the assurance of long-term environmental data to support their development and sustainability in the future. The loss or lack of continuous coverage for key satellite-derived environmental variables would reduce the ability of near-real-time habitat-based tools to be developed or to operate.

(iv)   Biological Realism Versus Statistical Model Fit

Generally, habitat-preference models are used to investigate relationships between animals and the environment, and to identify which environmental factors are statistically significant. More recently, there has been an increase in the use of statistical habitat-based models to predict or forecast occurrence or density of animals (Becker *et al.* 2012a, Forney *et al.* 2012, Pardo *et al.* 2015). However, we found, as we conducted traditional model-fit approaches on our blue whale habitat models and then reviewed the output predictions, that the statistical goodness of fit (e.g. Akaike Information Criterion) did not always equate to biologically realistic predictions. It is important to accurately identify the form of the relationship between the environmental variables and the species of interest, but unusual or spurious values in the new environmental data input into the model can lead to unexpected results in the output prediction maps. An expert review of the mapped predictions was an essential part of the process, to ensure that the model was generating realistic species distributions. Independent data sources, such as sightings data from the SWFSC line-transect surveys (Becker *et al.* 2012b) and photo-identification surveys (Calambokidis *et al.* 2009), played a key role in informing us what the expected seasonal distribution should look like. They were also an important data source for evaluating the performance of the telemetry-based habitat model and identifying strengths and limitations amongst all of the datasets. A comparison of our model predictions with the SWFSC line-transect survey sightings in 2005 and 2008 indicated fair agreement, based on the area-under-the-curve cross-validation statistics, calculated from receiver-operating-characteristic curves (Hazen *et al.* 2016).

Population size estimates from mark–recapture models (Calambokidis and Barlow 2013) were used to estimate absolute densities from our blue whale habitat model. We followed a similar approach to Aarts *et al.* (2008) in which the estimates of usage were scaled by the population abundance, to give absolute densities of animals (Hazen *et al.* 2016). Estimates of population abundance derived from other sources are therefore currently necessary in order for telemetry data to be processed in a way that will provide absolute density estimates. The monthly-output resolution of our predictions also required population abundance estimates for that region at a relatively high temporal scale. Although blue whale abundance estimates off the US west coast were available for the summer/autumn season, they were not available at the monthly, year-round resolution of our model predictions. We therefore used the monthly proportion of tracking positions in the waters off the US west coast to estimate the proportion of the population abundance that was expected to occur in that area for each month. The resulting density estimates were comparable to other model estimates (Forney *et al.* 2012).

Although our telemetry dataset was sufficiently large that it spanned all months of the year, it became clear during our modelling efforts that relationships with environmental variables were not consistent over the entire year. A single, year-round model overestimated the occurrence of blue whales in US west coast waters during the winter compared to information from the tracking and sightings data. The final models therefore included separate models for winter–spring (December–June) and summer–autumn (July–November). The two seasonal models contained the same statistically significant environmental variables, but the form of the relationships was different (Hazen *et al.* 2016). This likely reflects the animals exhibiting different behaviours according to the ocean conditions at different times of year, perhaps in response to changes in prey type, prey–environmental relationships, or other resource needs, such as nursery habitat. This has similarly been found for different populations of other marine species (Bailey *et al.* 2012). Different individuals or populations of a species may use different foraging strategies because of the specific prey species available in their geographical area or niche, or to reduce competition (Saulitis *et al.* 2000, Tremblay and Cherel 2003, Sargaent *et al.* 2007). Changes in resource needs occur seasonally, and across life stages that may also affect habitat preferences (Page *et al.* 2006, Shillinger *et al.* 2012). The transferability of a habitat model to other locations and populations would therefore require evaluation (e.g. Monsarrat *et al.* 2015).

(v)   Validation: What is the Truth?

It is important to evaluate and validate models to determine their performance and accuracy. One of the common practices has been to withhold a portion of the data as a test dataset (Guisan and Zimmermann 2000). However, this reduces the amount of data used in the model development, and the test dataset will still retain any biases that were in the original dataset, such as if the tracking data are mostly from animals tagged at the same location, or of the same sex or age group. It should be recognised that no method is perfect and each approach has its own benefits and limitations.

We found that model evaluation by comparison with independent datasets and models, such as those derived from the SWFSC shipboard surveys, provided a useful indicator of where there are consistencies or discrepancies, which helped to focus further analysis and research efforts (Brookes *et al.* 2013, Thompson *et al.* 2015). For example, we had relatively high estimates of the probability of occurrence in the northern portion of our study area, which was not supported by the tracking data. However, blue whales were caught off Alaska in commercial whaling, and identification photographs showed that 15 blue whales that were sighted off British Columbia and in the Gulf of Alaska had previously occurred off California (Calambokidis *et al.* 2009). It is possible that a suitable habitat may exist in these areas, as our models predict, but that blue whales have not yet fully re-established their traditional migration pattern and use of those habitats. The SWFSC also had more sightings offshore than we predicted (Hazen *et al.* 2016), which indicates that the blue whales tagged close to shore may not fully represent the population's distribution range. Further tagging or survey efforts offshore may help to determine the relative importance of this habitat and whether it is underestimated in our model.

Marine mammal distribution can be studied through visual sightings from land, boat, or aerial platforms, towed or static passive acoustic monitoring, and telemetry. The current challenge is developing approaches for incorporating these different data sources, which will ultimately lead to a better understanding of species distributions and movements. The SWFSC is currently involved in efforts to combine models of blue whale distribution from multiple data sources, including the WhaleWatch model.

(vi) Automation and Sustainability

The future sustainability of a near-real-time tool beyond the initial project development is most likely to be successful if the procedure is automated, so that as little human intervention as possible, and consequently funding, is necessary. Long-term support and management of the tool is probably best served by institutions with relatively steady funding, such as a government agency, as academics and industry tend to run on project funding cycles of 3–5 years, at most. If the same agency that is using the tool also manages it, any potential loss of connection with the end-user can be prevented, if there is a change of personnel at the institution that assisted with the tool development.

During the modelling process, a large amount of environmental data needed to be downloaded and formatted for the 104 blue whale tracks in the analysis, and the corresponding 200 simulated tracks per whale track, to obtain the case points for the model. For each of these thousands of positions, corresponding values for 13 environmental variables were downloaded. We learned during this process that it was important to carefully note exactly which environmental products were used in the model development, and to pass that information along to the team creating the automation code.

Scientists sometimes have difficulty reaching an end point and a final model, as there are always new ways to make tweaks, refinements, and improvements. However, the need to keep to schedule and complete the project in a timely manner required that we had to reach a point where the team agreed on the model and its predictive capabilities, so the personnel creating the automation code could then work with a final model version. Clear communication and documentation between the model developers and the personnel working on the automation process was much more important than we initially realised. The devil is in the details! Although the transition between development and operation seemed straightforward, the many questions involved in automating the code made us realise the importance of very clear and careful documentation, which will also be valuable if any updates are necessary in the future.

In our WhaleWatch project, the model processing is now conducted and the outputs are stored and hosted by the SWFSC Environmental Research Division. Redundancies, such as multiple calls to different sources of environmental data, can be included in the automation procedure. For example, our code includes downloading cholorophyll-*a* concentration data from the Moderate Resolution Imaging Spectroradiometer (MODIS), but when these are no longer available (as the instrument has already exceeded its expected lifespan), the system will instead download the data from the Visible Infrared Imaging Radiometer Suite (VIIRS). There have already been occasions when technical oversight and support was required when updates to the system were necessary. For example, when the SSH data was transferred from AVISO to CMEMS in May 2016, the SWFSC Environmental Research Division modified the procedure to download the data from this new site. This highlights the importance of operational funding, as this is always a significant effort that tends to be underestimated in most research proposals.

In addition to changes to the location of data sources, the data products themselves can also change over time. During the development of WhaleWatch, AVISO modified the time period over which the absolute dynamic topography was calculated. In version 15.0, the reference period of the SSH anomalies is now based on a 20-year period (1993–2012) whereas in the former version it was based on a 7-year period (1993–1999). This resulted in changes to the SSH-anomaly products. The subsequent effect on the predicted whale densities was relatively minor. However, we had not anticipated such changes would occur within the time period of developing the habitat-preference model, and the additional analysis required more time than we had anticipated. We would recommend including additional time in a project plan to deal with unexpected issues, and also for the review and evaluation process, so that team members and external researchers can be fully engaged and consulted across all phases of the project.

It should be noted that although 'near real time' refers to a short time delay, generally introduced by data transmission or processing, there is no specification of that delay time-period. The remotely sensed environmental data that are considered near real time may have a delay on the order of days or weeks. The delay for the monthly products that we have been using in WhaleWatch is 15 days for SST and chlorophyll-a concentration and 30 days for SSH. This delay could be reduced by using daily or weekly products, but some of these variables will suffer from missing data due to the presence of clouds, amongst other issues, which results in undesirable gaps in the prediction maps, particularly close to the coast. We had considerable discussion amongst the team on the most appropriate temporal resolution to use, given the trade-offs. It was difficult to come to a consensus amongst the team and, retrospectively, it would have been helpful to consider this earlier in the project development. Generally, it is recommended to use the same temporal resolution for the model development and predictions. A better model performance may be achieved through using finer-temporal-resolution environmental data, which more accurately describe the ocean conditions experienced by the animal, but gaps in the environmental data coverage may lead to more missing values and this reduces the sample size. The spatial resolution (25 × 25-km) of our model output was set to match that of previous habitat-based cetacean models by the SWFSC (e.g. Forney *et al.* 2012). The temporal resolution (monthly) was based on the minimum resolution that tended to have complete environmental information, with minimal gaps across our entire study area, for all seasons and for all of the environmental variables of interest. In the future, the delay of any environmental products used in prediction could be accounted for by including a lag time for the environmental variables in the model. For example, instead of integrating the environmental variables concurrently with whale track positions, the environmental data could lag behind the timing of the whale position, to determine if the earlier environmental conditions can be used as leading indicators of the distribution of the whales. Environmental data for March, for example, could be entered into the model to predict the whales' occurrence in April. The inclusion of parameters from ocean models, such as the Regional Ocean Modeling System (ROMS), may also help to fill gaps in temporal or spatial data, and to predict and forecast whale distributions, particularly for sub-surface variables (Becker *et al.* 2012a, Becker *et al.* 2016).

Future changes in the ocean conditions that occur as a result of climate change and other processes are likely to affect prey distributions and, consequently, top predators (Cuddington *et al.* 2013, Silber *et al.* 2016, Lefevre *et al.* 2017). Although our blue whale tracking dataset spanned 15 years, there may be environmental conditions experienced in the future that are not represented in this dataset. Consequently, the model predictions may increasingly lose accuracy if there are changes to the blue whale habitat that cause them to shift, expand, or contract their range (Hazen *et al.* 2012). Human activities, such as anthropogenic noise, could also play a role in affecting blue whale behaviour and movements (Goldbogen *et al.* 2013). The collection of new telemetry or other data sources, such as sightings or passive acoustic monitoring, is therefore required, to determine if the accuracy of the predictions changes over time and to identify when there is a sufficient decline in predictive ability that modification of the model is necessary.

8.4 LIMITATIONS AND UNCERTAINTIES

As stated in statistics, 'All models are wrong but some are useful' (Box 1979). A key question is, therefore, whether the model is sufficiently useful that it can benefit decision-making and natural-resource management (Forney *et al.* 2015). The accuracy of habitat-based models will depend on the quality of data input and the persistence of species–environment relationships. The blue whale data in our study were from animals tagged with Argos satellite transmitters (Mate *et al.* 1999, Bailey *et al.* 2009, Irvine *et al.* 2014). Argos satellite locations are typically of lower quality and accuracy (on the order of 1–10-km) for species that are above the water surface for short time intervals at sea, such as marine mammals (Costa *et al.* 2010). The location accuracy is improved by using GPS technologies (generally accurate to within 100 m, e.g. Mate *et al.* 2016), but the tagging duration may be reduced or duty-cycling may be required, as a result of the higher power requirements. The majority of blue whales tracked in our study were tagged off the coast of California and demonstrated relatively high site fidelity (Irvine *et al.* 2014). Individuals occurring further offshore may, therefore, have been under-represented in our telemetry dataset, as was indicated by comparisons with the SWFSC shipboard sightings (Hazen *et al.* 2016).

A key assumption of habitat-based models is that the established species–environment relationships will continue to persist in the future. However, these relationships may alter as a result of changes in the prey–environment relationships that can occur in response to anomalous conditions caused by climate change and other oceanographic processes. For example, there was an increase in the diversity of cetaceans occurring in Monterey Bay, California, during the 1997–1998 El Niño (Benson *et al.* 2002). This area appeared to serve as a refuge during this period of generally low prey availability. The anomalously warm conditions in 2015 were predicted to result in low densities of blue whales off the US west coast (Hazen *et al.* 2016), but there are few data available to test this prediction. Continued whale data collection is necessary to validate whether this did indeed lead to a reduction in the occurrence of blue whales or if it was indicative of a breakdown in the functional relationships in the model because of the extreme ocean conditions (Becker *et al.* 2014).

On the WhaleWatch webpage, the lower and upper confidence intervals for our predictions of the probability of blue whale occurrence are shown to indicate the uncertainty in our model estimates. However, there is an additional source of uncertainty in the calculation of the density estimates. This calculation uses population abundance estimates from independent data sources (Calambokidis and Barlow 2013) and also requires the estimation of the proportion of the population occurring within our study area, the US Exclusive Economic Zone, and how it varies monthly. Since monthly population abundance estimates were not available, we used the tracking data to estimate the proportion of the population occurring in the study area monthly. We do not know how this might vary in relation to ocean conditions and this is not taken into account in our current model.

Although the WhaleWatch tool provides a useful source of information for managers and other stakeholders on the year-round occurrence and densities of blue whales, which they did not have before, it is important that the limitations and uncertainties of the predictions are recognised. We recommend that users consider the confidence-interval range provided, rather than just the average prediction estimate, in their decision-making. Further data collection is also required to ensure that the level of model accuracy is maintained in the future and that the model is updated as necessary. In this way, WhaleWatch can continue to serve as a useful tool in the future.

8.5 CONCLUSIONS

A key lesson learned during the WhaleWatch project was the benefit of strong communication and clear definitions of roles amongst members of the team and with managers at NOAA. Since every team member was working on multiple projects at the time, and were located across four time zones, having regular conference calls, presentations of preliminary results to the team, and short-term deadlines combined with longer-term milestones were important for maintaining progress, overcoming issues, and completing the project on schedule.

The previous experience of the SWFSC Environmental Research Division of hosting and analysing data from the Tagging of Pacific Predators (TOPP) project (www.topp.org) on a live access server was extremely valuable for the WhaleWatch project. The experience and skills from WhaleWatch are now being applied in another project, EcoCast (Mazen *et al.* in press), which is developing a new fishery management tool that will predict, in near real time, the spatial distributions of important, highly migratory, ocean species, including non-target species (such as California sea lions, *Zalophus californianus*) and target catch (e.g. swordfish, *Xiphias gladius*) using telemetry and fisheries catch data, integrated with satellite-derived environmental data. These efforts illustrate the evolution of dynamic ocean management from single-species, single-environmental-variable tools (TurtleWatch), to single-species, multiple-environmental-variable models (WhaleWatch), to the multi-species, multiple-environmental--variable approach of EcoCast.

Habitat-based models rely on the inference that relationships between animals and their environment that are observed in the past will persist in the future. However, this may not be true if changes in climate result in increasingly anomalous ocean conditions. For example, in the northeast Pacific Ocean, there was a region of exceptionally warm water that extended to the continental shelf, in 2014–2015, and this anomalous event was followed by a strong El Niño (Bond *et al.* 2015, Di Lorenzo and Mantua 2016, Jacox *et al.* 2016). If marine heatwaves occur more regularly in the future and change ecological relationships (Hobday *et al.* 2016), the performance of the habitat model may decline. Future collection of whale data and model validation will be necessary to evaluate the accuracy of the WhaleWatch model predictions, and to update the habitat model if the predictive capability declines significantly. Combining telemetry and systematic survey data with information from citizen scientists, such as through sighting reports on the Whale Alert network (www.whalealert.org) and other opportunistic sightings, can enhance data collection. Efforts are also underway, as part of an International Whaling Commission's scientific committee, to develop ensemble models that combine blue whale models from multiple data sources to account for sampling biases, and to improve the temporal and spatial resolution of predictions, as a further aid to reducing human impacts. Integrating additional information, such as competition and other biotic interactions, may also help to improve the predictive capacity of the models (Guisan and Thuiller 2005, Silber *et al.* 2016). Other whale and protected species should also be considered, to ensure that there are not unintended consequences of management actions on other protected species (e.g. a reduced risk of ship strikes for blue whales, but an increased risk for humpback whales) (Redfern *et al.* 2013).

Sustained, long-term satellite programmes play a critical role for providing consistent, science-quality environmental data, which form the foundation of the habitat models used in near-real-time tools. For example, the NOAA Coral Reef Watch (https://coralreefwatch.noaa.gov) programme provides continuous monitoring of remotely sensed SST, which can be used to determine when bleaching response plans and appropriate management action should be initiated (Hughes *et al.* 2017). Environmental data from satellites, *in situ* measurements, and ocean models are essential for understanding the biological impacts of changes in ocean conditions and responding appropriately to mitigate them where necessary. The role of technology in tag development, remotely sensed data, and analysis capabilities will continue to increase the efficacy of these approaches, as the big data revolution transitions into marine resource management problem-solving.

Ultimately, the goal of WhaleWatch was to provide a scientific basis for management and other stakeholders to make informed decisions that could help reduce human impacts on whale populations. Blue whales are currently listed as Endangered on the IUCN Red List and under the US Endangered Species Act and are also protected under the US Marine Mammal Protection Act. Under US regulations, the PBR has been estimated at 2.3 blue whales per year for the eastern North Pacific population (Carretta *et al.* 2016). Given the small number of reported ship strikes, which likely greatly underestimates the true number, and the small PBR for blue whales, it will be difficult to assess any trends in ship collisions, and whether or not the WhaleWatch project has helped to reduce them. However, NOAA can monitor how the tool has affected their decisions, and whether it has improved their ability to initiate additional mitigation measures when necessary, such as vessel speed reductions or efforts to reduce the overlap between areas with high densities of whales and ship traffic. An important component of the effectiveness of these measures will be outreach and education, as well as enforcement (Silber *et al.* 2014, 2015). Similar efforts will also be necessary to reduce entanglements in fishing gear, which is becoming an increasingly important issue for humpback and blue whales off the US west coast (National Marine Fisheries Service 2017). In order to reduce human–wildlife conflicts, near-real-time tools and adaptive management approaches, such as dynamic ocean management, will be increasingly vital to allow continued growth and survival of marine mammal populations and human use of our shared ocean resources.

ACKNOWLEDGEMENTS

Funding for the WhaleWatch project was provided under the interagency NASA, United States Geological Survey, National Park Service, US Fish and Wildlife Service, Smithsonian Institution Climate and Biological Response program, Grant Number NNX11AP71G and NOAA's Integrated Ecosystem Assessment program. Tag deployments were funded through Oregon State University's Marine Mammal Institute Endowed Program. We thank the many people who assisted with tagging and tag development from the Marine Mammal Institute as well as the various crews of the R/V *Pacific Storm*. The National Marine Fisheries Service authorised this research under permit numbers 841 (1993–1998), 369–1440 (1999 – 2004), and 369–1757 (2005 – 2010) to Dr Bruce Mate. Tagging in Mexican waters was conducted under permits issued by the Secretaría de Medio Ambiente y Recursos Naturales, Mexico (permit number DOO 02.8319 and SGPA/DGVS 0576). All research was approved by the Oregon State University Animal Care and Use Committee. Thank you to Justin Peters, Penny Ruvelas, Michael Milstein, Laura Oremland, and Avi Litwack at the NOAA for their help and support of the project. We also thank Dave Foley and Dale Robinson for their assistance with the satellite-derived environmental data.

REFERENCES

Aarts, G., M. MacKenzie, B. McConnell, M. Fedak, and J. Matthiopoulos (2008). Estimating space-use and habitat preference from wildlife telemetry data. *Ecography*, **31**, 140–160.

Aarts, G., J. Fieberg, and J. Matthiopoulos (2012). Comparative interpretation of count, presence–absence and point methods for species distribution models. *Methods in Ecology and Evolution*, **3**, 177–187.

Ainley, D. G., D. Jongsomjit, G. Ballard, *et al.* (2012). Modeling the relationship of Antarctic minke whales to major ocean boundaries. *Polar Biology*, **35**, 281–290.

Bailey, H., G. Shillinger, D. Palacios, *et al.* (2008). Identifying and comparing phases of movement by leatherback turtles using state-space models. *Journal of Experimental Marine Biology and Ecology*, **356**, 128–135.

Bailey, H., B. R. Mate, D. M. Palacios, *et al.* (2009). Behavioural estimation of blue whale movements in the northeast Pacific from state-space model analysis of satellite tracks. *Endangered Species Research*, **10**, 93–106.

Bailey, H., S. R. Benson, G. L. Shillinger, *et al.* (2012). Identification of distinct movement patterns in Pacific leatherback turtle populations influenced by ocean conditions. *Ecological Applications*, **22**, 735–747.

Bailey, H., P. S. Hammond, and P. M. Thompson (2014). Modelling harbour seal habitat by combining data from multiple tracking systems. *Journal of Experimental Marine Biology and Ecology*, **450**, 30–39.

Barlow, J. and K. A. Forney (2007). Abundance and population density of cetaceans in the California Current ecosystem. *Fishery Bulletin*, **105**, 509–526.

Baumgartner, M. F., D. M. Fratantoni, T. P. Hurst, *et al.* (2013). Real-time reporting of baleen whale passive acoustic detections from ocean gliders. *Journal of the Acoustical Society of America*, **134**, 1814–1823.

Becker, J. J., D. T. Sandwell, W. H. F. Smith, *et al.* (2009). Global bathymetry and elevation data at 30 arc seconds resolution: SRTM30_PLUS, revised for Marine Geodesy, January 20, 2009.

Becker, E. A., K. A. Forney, M. C. Ferguson, *et al.* (2010). Comparing California Current cetacean-habitat models developed using *in situ* and remotely sensed sea surface temperature data. *Marine Ecology Progress Series*, **413**, 163–183.

Becker, E. A., D. G. Foley, K. A. Forney, *et al.* (2012a). Forecasting cetacean abundance patterns to enhance management decisions. *Endangered Species Research*, **16**, 97–112.

Becker, E. A., K. A. Forney, M. C. Ferguson, J. Barlow, and J. V. Redfern (2012b). Predictive modeling of cetacean densities in the California Current Ecosystem based on summer/fall ship surveys in 1991–2008. US Department of Commerce, NOAA Technical Memorandum NMFS-SWFSC-499.

Becker, E. A., K. A. Forney, D. G. Foley, *et al.* (2014). Predicting seasonal density patterns of California cetaceans based on habitat models. *Endangered Species Research*, **23**, 1–22.

Becker, E. A., K. A. Forney, P. C. Fiedler, *et al.* (2016). Moving towards dynamic ocean management: How well do modeled ocean products predict species distributions? *Remote Sensing*, **8**, 149.

Benson, S. R., D. A. Croll, B. B. Marinovic, F. P. Chavez, and J. T. Harvey (2002). Changes in the cetacean assemblage of a coastal upwelling ecosystem during El Niño 1997–98 and La Niña 1999. *Progress in Oceanography*, **54**, 279–291.

Berman-Kowalewski, M., F. M. D. Gulland, S. Wilkin, *et al.* (2010). Association between blue whale (*Balaenoptera musculus*) mortality and ship strikes along the California coast. *Aquatic Mammals*, **36**, 59–66.

Best, B. D., P. N. Halpin, A. J. Read, *et al.* (2012). Online cetacean habitat modeling system for the US east coast and Gulf of Mexico. *Endangered Species Research*, **18**, 1–15.

Bond, N. A., M. F. Cronin, H. Freeland, and N. Mantua (2015). Causes and impacts of the 2014 warm anomaly in the NE Pacific. *Geophysical Research Letters*, **42**, 3414–3420.

Box, G. E. P. (1979). Robustness in the strategy of scientific model building. In R. L. Launer and G. N. Wilkinson, eds. *Robustness in Statistics*, New York, NY: Academic Press, pp. 201–236.

Brookes, K. L., H. Bailey, and P. M. Thompson (2013). Predictions from harbor porpoise habitat association models are confirmed by long-term passive acoustic monitoring. *Journal of the Acoustical Society of America*, **134**, 2523–2533.

Brown, M. W., S. D. Kraus, C. K. Slay, and L. P. Garrison (2007). Surveying for discovery, science, and management. In S. D. Kraus and R. M. Rolland, eds., *The*

Urban Whale: North Atlantic Right Whales at the Crossroads, Cambridge, Massachusetts: Harvard University Press, pp. 105–137.

Calambokidis, J. and J. Barlow (2004). Abundance of blue and humpback whales in the eastern North Pacific estimated by capture–recapture and line-transect methods. *Marine Mammal Science*, **20**, 63–85.

Calambokidis, J. and J. Barlow (2013). Updated abundance estimates of blue and humpback whales off the US West Coast incorporating photo-identifications from 2010 and 2011. Cascadia Research final report for contract AB133F-10-RP-0106.

Calambokidis, J., G. S. Schorr, G. H. Steiger, *et al.* (2008). Insights into the underwater diving, feeding, and calling behavior of blue whales from a suction-cup-attached video-imaging tag (Crittercam). *Marine Technology Society Journal*, **41**, 19–29.

Calambokidis, J., J. Barlow, J. K. B. Ford, T. E. Chandler, and A. B. Douglas (2009). Insights into the population structure of blue whales in the eastern North Pacific from recent sightings and photographic identification. *Marine Mammal Science*, **25**, 816–832.

Carretta, J. V., S. M. Wilkin, M. M. Muto, and K. Wilkinson (2013). Sources of human-related injury and mortality for U.S. Pacific west coast marine mammal stock assessments, 2007–2011. US Department of Commerce, NOAA Technical Memorandum NOAA-TM-NMFS-SWFSC-514.

Carretta, J. V., E. M. Oleson, J. Baker, *et al.* (2016). U.S. Pacific marine mammal stock assessments: 2015. US Department of Commerce, NOAA Technical Memorandum NOAA-TM-NMFS-SWFSC-561.

Casey, K. S., T. B. Brandon, P. Cornillon, and R. Evans (2010). The past, present and future of the AVHRR Pathfinder SST Program. In V. Barale, J. F. R. Gower, and L. Alberotanza, eds. *Oceanography from Space: Revisited*. Springer, pp. 323–341. See https://pathfinder.nodc.noaa.gov/OFS_21_Cas_09Dec2009.pdf.

Clapham, P. J., S. B. Young, and R. L. Brownell Jr (1999). Baleen whales: conservation issues and the status of the most endangered populations. *Mammal Review*, **29**, 35–60.

Costa, D. P., P. W. Robinson, J. P. Y. Arnould, *et al.* (2010). Accuracy of Argos locations of pinnipeds at-sea estimated using Fastloc GPS. *PLOS ONE*, **5**, e8677,.

Croll, D. A., B. Marinovic, S. Benson, F. P. *et al.* (2005). From wind to whales: trophic links in a coastal upwelling system. *Marine Ecology Progress Series*, **289**, 117–130.

Cuddington, K., M. J. Fortin, L. R. Gerber, *et al.* (2013). Process-based models are required to manage ecological systems in a changing world. *Ecosphere*, 4, article 20.

Di Iorio, L. and C. W. Clark (2010). Exposure to seismic survey alters blue whale acoustic communication. *Biology Letters*, **6**, 51–54.

Di Lorenzo, E. and N. Mantua (2016). Multi-year persistence of the 2014/15 North Pacific marine heatwave. *Nature Climate Change*, **6**, 1042–1047.

Edrén, S. M. C., M. S. Wisz, J. Teilmann, R. Dietz, and J. Söderkvist (2010). Modelling spatial patterns in harbour porpoise satellite telemetry data using maximum entropy. *Ecography*, **33**, 698–708.

Fiedler, P. C., S. B. Reilly, R. P. Hewitt, *et al.* (1998). Blue whale habitat and prey in the California Channel Islands. *Deep-Sea Research II*, **45**, 1781–1801.

Forney, K. A., M. C. Ferguson, E. A. Becker, *et al.* (2012). Habitat-based spatial models of cetacean density in the eastern Pacific Ocean. *Endangered Species Research*, **16**, 113–133.

Forney, K. A., E. A. Becker, D. G. Foley, J. Barlow, and E. M. Oleson (2015). Habitat-based models of cetacean density and distribution in the central North Pacific. *Endangered Species Research*, **27**, 1–20.

Godley, B. J. and R. P. Wilson (2008). Tracking vertebrates for conservation: introduction. *Endangered Species Research*, **4**, 1–2.

Goldbogen, J. A., B. L. Southall, S. L. DeRuiter, *et al.* (2013). Blue whales respond to simulated mid-frequency military sonar. *Proceedings of the Royal Society B*, **280**, 20130657.

Guisan, A. and N. E. Zimmermann (2000). Predictive habitat distribution models in ecology. *Ecological Modelling*, **135**, 147–186.

Guisan, A. and W. Thuiller (2005). Predicting species distribution: offering more than simple habitat models. *Ecology Letters*, **8**, 993–1009.

Hays, G. C., L. C. Ferreira, A. M. M. Sequeira, *et al.* (2016). Key questions in marine megafauna movement ecology. *Trends in Ecology & Evolution*, **31**, 463–475.

Hazen, E. L., S. Jorgensen, R. R. Rykaczewski, *et al.* (2012). Predicted habitat shifts of Pacific top predators in a changing climate. *Nature Climate Change*, **3**, 234–238.

Hazen, E. L., D. Palacios, K. A. Forney, *et al.* (2016). WhaleWatch: a dynamic management tool for predicting blue whale density in the California Current. *Journal of Applied Ecology*, doi: 10.1111/1365-2664.12820.

Hazen, E. L., K. Scales, S. M. Maxwell, *et al.* A dynamic ocean management tool to reduce by catch and support sustainable fisheries. *Scientific Advances*. In press.

Hobday, A. J., J. R. Hartog, C. M. Spillman, and O. Alves (2011). Seasonal forecasting of tuna habitat for dynamic spatial management. *Canadian Journal of Fisheries and Aquatic Sciences*, **68**, 898–911.

Hobday, A. J., S. M. Maxwell, J. Forgie, *et al.* (2014). Dynamic ocean management: integrating scientific and technological capacity with law, policy, and management. *Stanford Environmental Law Journal*, **33**, 125–165.

Hobday, A. J., L. V. Alexander, S. E. Perkins, *et al.* (2016). A hierarchical approach to defining marine heatwaves. *Progress in Oceanography*, **141**, 227–238.

Horning, N., J. A. Robinson, E. J. Sterling, W. Turner, and S. Spector (2010). *Remote Sensing for Ecology and Conservation: A Handbook of Techniques.* Oxford: Oxford University Press.

Howell, E. A., D. R. Kobayashi, D. M. Parker, G. H. Balazs, and J. J. Polovina (2008). TurtleWatch: a tool to aid in the bycatch reduction of loggerhead turtles *Caretta caretta* in the Hawaii-based pelagic longline fishery. *Endangered Species Research*, **5**, 267–278.

Howell, E. A., A. Hoover, S. R. Benson, *et al.* (2015). Enhancing the TurtleWatch product for leatherback sea turtles, a dynamic habitat model for ecosystem-based management. *Fisheries Oceanography*, **24**, 57–68.

Hughes, T. P., J. T. Kerry, M. Álvarez-Noriega, *et al.* (2017). Global warming and recurrent mass bleaching of corals. *Nature*, **543**, 373–377.

Irvine, L. M., B. R. Mate, M. H. Winsor, *et al.* (2014). Spatial and temporal occurrence of blue whales off the US west coast, with implications for management. *PLOS ONE*, **9**, e102959.

Jacox, M. G., E. L. Hazen, K. D. Zaba, *et al.* (2016). Impacts of the 2015–2016 El Niño on the California Current System: early assessment and comparison to past events. *Geophysical Research Letters*, **43**, 7072–7080.

Jonsen, I. D., J. M. Fleming, and R. A. Myers (2005). Robust state-space modeling of animal movement data. *Ecology*, **86**, 2874–2880.

Jonsen, I. D., M. Basson, S. Bestley, *et al.* (2013). State-space models for bio-loggers: a methodological road map. *Deep-Sea Research II*, **88–89**, 34–46.

Kraus, S. D., M. W. Brown, H. Caswell, *et al.* (2005). North Atlantic right whales in crisis. *Science*, **309**, 561–562.

Laist, D. W., A. R. Knowlton, J. G. Mead, A. S. Collet, and M. Podesta (2001). Collisions between ships and whales. *Marine Mammal Science*, **17**, 35–75.

Laist, D. W., A. R. Knowlton, and D. Pendleton (2014). Effectiveness of mandatory vessel speed limits for protecting North Atlantic right whales. *Endangered Species Research*, **23**, 133–147.

Lefevre, S., D. J. McKenzie, and G. E. Nilsson (2017). Models projecting the fate of fish populations under climate change need to be based on valid physiological mechanisms. *Global Change Biology*, doi: 10.1111/gcb.13652.

Lewison, R., A. J. Hobday, S. Maxwell, *et al.* (2015). Dynamic ocean management: identifying the critical ingredients of dynamic approaches to ocean resource management. *BioScience*, **65**, 486–498.

Mate, B., B. Lagerquist, and J. Calambokidis (1999). Movements of North Pacific blue whales during the feeding season off southern California and their southern fall migration. *Marine Mammal Science*, **15**, 1246–1257.

Mate, B., R. Mesecar, and B. Lagerquist (2007). The evolution of satellite-monitored radio tags for large whales: one laboratory's experience. *Deep-Sea Research II*, **54**, 224–247.

Mate, B. R., L. M. Irvine, and D. M. Palacios (2016). The development of an intermediate-duration tag to characterize the diving behavior of large whales. *Ecology and Evolution*, doi: 10.1002/ece3.2649.

Maxwell, S. M., E. L. Hazen, R. L. Lewison, *et al.* (2015). Dynamic ocean management: defining and conceptualizing real-time management of the ocean. *Marine Policy*, **58**, 42–50.

Monsarrat, S., M. G. Pennino, T. D. Smith, *et al.* (2015). Historical summer distribution of the endangered North Atlantic right whale (*Eubalaena glacialis*): a hypothesis based on environmental preferences of a congeneric species. *Diversity and Distributions*, doi: 10.1111/ddi.12314.

Moore, J. E., B. P. Wallace, R. L. Lewison, *et al.* (2009). A review of marine mammals, sea turtle and seabird bycatch in USA fisheries and the role of policy in shaping management. *Marine Policy*, **33**, 435–451.

National Geophysical Data Center (2006). 2-minute gridded global relief data (ETOPO2) v2. National Geophysical Data Center, NOAA. doi: 10.7289/V5J1012Q (accessed May 2015).

National Marine Fisheries Service (2017). 2016 West coast entanglement summary. US Department of Commerce, National Oceanic and Atmospheric Administration, National Marine Fisheries Service. See www.westcoast.fisher ies.noaa.gov/publications/protected_species/marine_mammals/cetaceans/wc r_2016_whale_entanglements_3-26-17_final.pdf. Accessed March 2017.

Nowacek, D. P., L. H. Thorne, D. W. Johnston, and P. L. Tyack (2007). Responses of cetaceans to anthropogenic noise. *Mammal Review*, **37**, 81–115.

O'Reilly, J. E., S. Maritorena, B. G. Mitchell, *et al.* (1998). Ocean color chlorophyll algorithms for SeaWiFS. *Journal of Geophysical Research*, **103**, 24,937–24,953.

Page, B., J. McKenzie, M. D. Sumner, M. Coyne, and S. D. Goldsworthy (2006). Spatial separation of foraging habitats among New Zealand fur seals. *Marine Ecology Progress Series*, **323**, 263–279.

Pardo, M. A., T. Gerrodette, E. Beier, *et al.* (2015). Inferring cetacean population densities from the absolute dynamic topography of the ocean in a hierarchical Bayesian framework. *PLOS ONE*, **10**, e0120727.

Parks, S. E., J. D. Warren, K. Stamiezkin, C. A. Mayo, and D. Wiley (2012). Dangerous dining: surface foraging of North Atlantic right whales increases risk of vessel collisions. *Biology Letters*, **8**, 57–60.

Patterson, T. A., L. Thomas, C. Wilcox, O. Ovaskainen, and J. Matthiopoulos (2008). State-space models of individual animal movement. *Trends in Ecology and Evolution*, **23**, 87–94.

Read, A. J., P. Drinker, and S. Northridge (2006). Bycatch of marine mammals in US and global fisheries. *Conservation Biology*, **20**, 163–169.

Redfern, J. V., M. C. Ferguson, E. A. Becker, *et al.* (2006). Techniques for cetacean-habitat modeling. *Marine Ecology Progress Series*, **310**, 271–295.

Redfern, J. V., M. F. McKenna, T. J. Moore, *et al.* (2013). Assessing the risk of ships striking large whales in marine spatial planning. *Conservation Biology*, **27**, 292–302.

Reilly, S. B., J. L. Bannister, P. B. Best, *et al.* (2008). *Balaenoptera musculus*. The IUCN Red List of Threatened Species 2008:e.T2477A9447146. www.iucnred list.org/pdflink.9447146. Accessed May 2017.

Reynolds, R. W., T. M. Smith, C. Liu, *et al.* (2007). Daily high-resolution-blended analyses for sea surface temperature. *Journal of Climate*, **20**, 5473–5496.

Rutz, C. and G. C. Hays (2009). New frontiers in biologging science. *Biology Letters*, **5**, 289–292.

Santora, J. A., W. J. Sydeman, I. D. Schroeder, B. K. Wells, and J. C. Field (2011). Mesoscale structure and oceanographic determinants of krill hotspots in the California Current: implications for trophic transfer and conservation. *Progress in Oceanography*, **91**, 397–409.

Santora, J. A., J. C. Field, I. D. Schroeder, *et al.* (2012). Spatial ecology of krill, micronekton and top predators in the central California Current: implications for defining ecologically important areas. *Progress in Oceanography*, **106**, 154–174.

Sargaent, B. L., A. J. Wirsing, M. R. Heithaus, and J. Mann (2007). Can environmental heterogeneity explain individual foraging variation in wild bottlenose dolphins (*Tursiops* sp.)? *Behavioral Ecology and Sociobiology*, **61**, 679–688.

Saulitis, E., C. Matkin, L. Barrett-Lennard, K. Heise, and G. Ellis (2000). Foraging strategies of sympatric killer whale (*Orcinus orca*) populations in Prince William Sound, Alaska. *Marine Mammal Science*, **16**, 94–109.

Sears, R. and W. F. Perrin (2009). Blue whale (*Balaenoptera musculus*). In W. F. Perrin, B. Würsig, and J. G. M. Thewissen, eds., *Encyclopedia of Marine Mammals*. San Diego, CA: Academic Press, pp. 120–124.

Shillinger, G. L., H. Bailey, S. J. Bograd, *et al.* (2012). Tagging through the stages: technical and ecological challenges in observing life histories through biologging. *Marine Ecology Progress Series*, **457**, 165–170.

Silber, G. K., J. D. Adams, and C. J. Fonnesbeck (2014). Compliance with vessel speed restrictions to protect North Atlantic right whales. *PeerJ*, **2**, e399.

Silber, G. K., J. D. Adams, M. J. Asaro, *et al.* (2015). The right whale mandatory ship reporting system: a retrospective. *PeerJ*, **3**, e866.

Silber, G. K., M. Lettrich, and P. O. Thomas (2016). Report of a workshop on best approaches and needs for projecting marine mammal distributions in a changing climate. 12–14 January 2016, Santa Cruz, California, USA. US Department of Commerce, NOAA Technical Memorandum NMFS-OPR-54.

Simons, R. A. (2016). ERDDAP. See https://coastwatch.pfeg.noaa.gov/erddap.

Stone, G., L. Florez-Gonzalez, and S. Katona (1990). Whale migration record. *Nature*, **346**, 705.

Thomas, P. I., R. R. Reeves, and R. L. Brownell Jr (2016). Status of the world's baleen whales *Marine Mammal Science*, **32**, 682–734.

Thompson, P. M., K. L. Brookes, and L. S. Cordes (2015). Integrating passive acoustic and visual data to model spatial patterns of occurrence in coastal dolphins. *ICES Journal of Marine Science*, **72**, 651–660.

Tremblay, Y. and Y. Cherel (2003). Geographic variation in the foraging behaviour, diet and chick growth of rockhopper penguins. *Marine Ecology Progress Series*, **251**, 279–297.

Van Parijs, S. M., C. W. Clark, R. S. Sousa-Lima, *et al.* (2009). Management and research applications of real-time and archival passive acoustic sensors over varying temporal and spatial scales. *Marine Ecology Progress Series*, **395**, 21–36.

Van Waerebeek, K., A. N. Baker, F. Félix, *et al.* (2007). Vessel collisions with small cetaceans worldwide and with large whales in the southern hemisphere, an initial assessment. *Latin American Journal of Aquatic Mammals*, **6**, 43–69.

Vanderlaan, A. S. M. and C. T. Taggart (2007). Vessel collisions with whales: the probability of lethal injury based on vessel speed. *Marine Mammal Science*, **23**, 144–156.

Wiley, D. N., M. A. Thompson, R. M. Pace, and J. Levenson (2011). Modeling speed restrictions to mitigate lethal collisions between ships and whales in the Stellwagen Bank National Marine Sanctuary, USA. *Biological Conservation*, **144**, 2377–2381.

9 The Evolution of Remote Sensing Applications Vital to Effective Biodiversity Conservation and Sustainable Development

Karyn Tabor and Jennifer Hewson

9.1 INTRODUCTION

This chapter considers how the conservation community has applied advances in satellite remote sensing over the last two decades and how this has transformed conservation practices and decision-making. The chapter is not a review of remote sensing across all conservation organisations. It is written from the perspective of conservation practitioners at Conservation International (CI), a US-based international conservation, non-governmental, organisation that aims to empower societies to protect nature for the benefit of people across the globe. CI has incorporated remote sensing technologies into its activities for over 20 years and, therefore, was an early adopter of the technology compared to many other conservation organisations. We describe how advances in remote sensing in the past two decades, whether by design or not, have influenced conservation practices. We highlight the conservation successes that these advances afforded, both in terms of improved land-use management and more efficient use of resources. Finally, we provide recommendations for focal areas, which we feel will lead to even greater conservation successes, and discuss advances on the horizon for satellite remote sensing that may lead to the next set of breakthroughs for advancing conservation science and applications.

In the late 1990s, remote sensing was applied by a few conservation practitioners with highly technical skill sets, working in organisations supported by computing environments that were capable of handling relatively complex data, and enabled by proprietary software. Very few conservation organisations employed remote sensing specialists at that time. CI represented a novelty in the conservation world, employing half a dozen remote sensing specialists. Today, satellite imagery is more readily accessible to many conservation practitioners, and much of it is available at no cost. Indeed, the general public can even view satellite imagery and perform basic data analyses using many online platforms. Access to satellite imagery within the broader conservation community escalated the use of remotely sensed data; today, the use of remotely sensed data to support core conservation applications, such as environmental monitoring, mapping, and priority setting, is almost compulsory.

The authors of this chapter pursued a technical career, utilising remote sensing early in their careers. They observed a steady growth in the use of remote sensing technologies by colleagues who took a forestry or ecology career path but who increasingly recognised the value of remote sensing in support of their activities. This reflects the increasing access to data, as well as the synoptic view and the potential for regular monitoring that are afforded by remote sensing technologies. Today, many other conservation and development organisations integrate remote sensing data and derived products into their initiatives. Examples include the World Resource Institute's Global Forest Watch (GFW) platform, the Center for International Forestry Research's Sustainable Wetlands Adaptation and Mitigation Program, and The Nature Conservancy's Ecoregions project, to name a few. At CI, we continue to explore the use of remote sensing data for a host of initiatives, such as integrated forest monitoring, near real-time response, non-forest habitat mapping, and blue carbon applications. Today, CI strives to adopt new technologies into its activities, in order to build and share tools that increase accessibility to, and facilitate use of, remote sensing data for conservation applications. This includes incorperating governmental and commercial tools that make satellite data available in cloud-computing environments, such as Google Earth Engine (GEE) and NASA's Earth Exchange (NEX) platform.

9.2 HISTORY OF REMOTE SENSING USE BY CONSERVATION ORGANISATIONS

Natural-Resource Mapping

Mapping exemplifies one of the earliest and most basic forms of data analysis; in fact, maps, in the form of landscape representations, were found on cave walls dating back to early human civilisation. Naturalresource maps were hand-drawn until the invention of photography. With aerial photography, conservation practitioners were able to combine aerial photos with field-survey information and generate landcover classifications. However, while aerial photos were available in countries with technological resources, the availability of such data was more limited for conservation applications in the tropics. This was primarily due to high costs and the logistics involved in acquiring such data. With the launch of the 70-m spatial resolution Multispectral Scanner (MSS) on board Landsat 1 in 1972, the conservation community, for the first time, could map extents and monitor conditions of different habitats. The Landsat programme's observational capability to distinguish multiple vegetation types proved vital to the conservation remote sensing community making it the 'workhorse' in applied remote sensing for conservation. The community began designing conservation actions that were informed by this synoptic and repeatable view of the landscape (De Wulf *et al.* 1988, Luque 2000, Steininger *et al.* 2001). Although routine global mapping at medium spatial resolutions would take another three decades to accomplish, the Landsat missions represented a new era in conservation remote sensing.

Several sensor design modifications, beginning with the Thematic Mapper on board Landsat 4 (see Chapter 2), enhanced the ability of conservation practitioners to map different habitats. The 30-m spatial resolution and 170-km × 186-km (~3 million hectares) footprint of these images were well suited to conservation applications. It proved ideal for land-cover and land-use mapping at the landscape scale, the scale at which practitioners seek to understand the interactions between human and ecological dimensions. Early land-cover mapping techniques employed by the conservation community using Landsat relied heavily on manual interpretation, which was very time-consuming. As computer-assisted mapping techniques emerged in the mid to late 1990s, conservation practitioners produced land-cover maps more efficiently over larger geographic areas. The introduction of algorithms in computer-assisted mapping techniques facilitated the analysis of multi-dimensional spectral information. This yielded outputs much more efficiently than an analyst visually interpreting an entire Landsat scene. However, many practitioners interested in using Landsat for habitat mapping and conservation priority setting were based in tropical countries and had limited technical capacity. Thus, the spread of such techniques progressed rather slowly.

Field information has always been crucial for informing satellite image interpretation and for assessing the accuracy of image classifications. At CI, we employed both aerial photography and videography along with field data (Figure 9.1), where available, typically generated by botanists or ecologists in our field offices or partner organisations (Box 9.1). However, digital field data were rare. Therefore, we shipped printed satellite-imagery maps to field experts to draw circles on vegetation stands and label known vegetation types. These maps were returned, marked up and smeared with mud, but the information proved extremely valuable as it could be used to create polygons, based on expert information, and in turn train computer algorithms to analyse the multispectral information and generate wall-to-wall image classifications. These data also informed accuracy assessments of our map

FIGURE 9.1 Conservation International's aerial survey programme in action. CI's remote sensing analyst is collecting aerial videography and photo sampling during a survey in Madagascar, ~2003. The geotagged imagery was later compiled into a GIS and used for validating a forest cover and change product. As a ground-truthing tool, aerial sampling allowed conservation practitioners to systematically gather data over a much larger area than field data collection alone. (Photo Credit: Daniel Juhn.)

BOX 9.1 **CI's aerial survey programme**

CI's aerial survey programme operated from the early 1990s through the early 2010s, collecting aerial photographs and videography to inform training data for land-cover classifications and for performing accuracy assessments (see Figure 9.1). Several constraints existed in conducting aerial surveys, including high costs, timing restrictions (e.g. we could not fly in the rainy season when it was too wet and cloudy), local zoning restrictions on flying, and risk of equipment failure. This was an expensive and time-intensive endeavour; however, it was necessary due to limited ground data or expert knowledge in a region.

---

classifications. Despite the time spent sending and receiving paper maps, this process was quicker than an analyst digitising land-cover types from a scene. It improved the efficiency of production and accuracy of land-cover maps. The advancement of methods to utilise panchromatic bands for higher spatial resolution, pan-sharpened, spectral images facilitated evaluations of map accuracy without the need for expensive high-resolution imagery or aerial surveys. This further improved the accuracy of the maps we produced for conservation applications. The continued evolution of improved sensor design for terrestrial monitoring is one reason for the prevailing use of Landsat and Satellite Pour l'Observation de la Terre (SPOT) satellites; the recently launched ESA Sentinel satellites now also contribute data for land-cover and land-use mapping applications.

The provision of remote sensing-derived land-cover and land-use mapping products has advanced conservation priority setting. Previous priority-setting activities had relied heavily on expert knowledge and hand-drawn maps that were created through participatory mapping exercises during priority-setting workshops. Examples include delineating protected-area boundaries, the identification of biodiversity conservation corridors, and the reassessment of species status on the International Union for Conservation of Nature (IUCN) Red List, based on the remaining area of suitable habitat and rates of habitat loss.

The mapping of ecosystem extent and condition can identify and assess the value that natural ecosystems provide for biodiversity, as well as ecosystem service provision for human well-being. Although not covered in this book, mapping of natural capital (i.e. ecosystems providing key services for carbon sequestration, clean air, and water regulation) represents a key step towards the quantification of the value of nature's assets. Including ecosystem values as 'natural capital' in national accounting enables countries to make smart decisions for sustainable development pathways. For example, CI led one of the frist demonstrations for valuing natural capital as part of the World Bank-led Wealth Accounting and the Valuation of Ecosystem Services (WAVES) initiative. The case study, located in Madagascar, relied on accurate, national-scale land-cover maps derived from Landsat and SPOT to account for the value of water, forest, and tourism. It revealed that half of Madagascar's assets are its natural capital (Portela *et al.* 2012).

Repeated Landscape Monitoring

The long-term monitoring of habitat status and trends is essential to inform land-use policies and conservation actions. In order to meet these needs, national space agencies have made strategic decisions within their organisations and with each other to develop missions that support the continuity of observations. Arguably, the most relied upon remotely sensed data for informing national and international policy are routine forest-cover monitoring data from Landsat. In fact, remote sensing-derived forest-cover and deforestation maps (e.g. Harper *et al.* 2007, Tabor *et al.* 2010) provided the backbone for global initiatives including the Global Conservation Fund (GCF) and the Critical Ecosystem Partnership Fund (CEPF; Box 9.2).

Forest cover and change products also play a critical role in many countries' commitments to reductions in carbon emissions,

---

BOX 9.2 **The GCF and the CEPF**

The GCF, which completed in late 2016, represented a multi-year programme to create and expand protected areas in biodiversity hotspots and key marine areas, and to perform monitoring of change in and around these investment sites. Much of the monitoring performed in support of GCF used remote sensing data, namely Landsat. The CEPF, a collaborative initiative involving L'Agence Française de Développement, the European Union, the Government of Japan, the World Bank, Conservation International, the GCF, and the MacArthur Foundation, provides grants to non-governmental organisations, community-based organisations, and local partners to develop initiatives to protect biodiversity hotspot areas and, in turn, support the provision of both local and global ecosystem services. Many of the indicators that CEPF employs to monitor the initiatives, including the change in habitat extent and the change in target species threat levels, are informed using remote sensing imagery and derived products.

---

and are essential to the carbon credit market. For example, the REDD+ (Reducing Emissions from Deforestation and Forest Degradation plus the enhancement of forest carbon stocks, sustainable management of forests, and conservation of forest carbon stocks) initiative aims to mitigate climate change through enhanced forest management. To participate in REDD+, countries are required to implement four elements, including the development of: a national strategy or action plan, a forest reference level or forest reference emissions level (FRL/FREL), a national forest monitoring system (NFMS), and a system for implementing safeguards. Both the development of the FRL and the NFMS generally depend heavily on remote sensing data and derived products and inform the generation of annual estimates of forest change and required biennial reporting. Standards for establishing carbon baselines, routine forest monitoring, and reporting require consistent, accurate, and verifiable methods with low-cost techniques, and the spatial and spectral resolutions of the Landsat and, most recently, the Sentinel sensors, are optimal for national-scale forest monitoring and low-cost validation. These data are also essential to project future trends in land-cover changes from the project level to the regional and national scale, based on a range of policy scenarios (Angelsen *et al.* 2009). In addition to informing land-use policy decisions, the data are equally key in evaluating the effectiveness of the policies. Impact evaluations of conservation investments rely heavily on maps of forest-cover extent over time to inform effective decision-making on the best conservation practices, meeting the desired outcomes of slowing deforestation rates (Gaveau *et al.* 2009).

In addition to moderate-resolution forest-cover monitoring, low-spatial-resolution satellites with daily global coverage are highly valuable for long-term global monitoring of ecosystem dynamics. For example, the research and applications community exploited the high revisit rate of the Advanced Very High Resolution Radiometer (AVHRR), a sensor designed to monitor clouds, to find windows through the clouds to view the ground in persistently cloudy regions. The development of the Normalised Difference Vegetation Index (NDVI) (Tucker 1979), using the repeated monitoring of AVHRR, represented a breakthrough for understanding seasonal dynamics and long-term trends, including timing of green-up and senescence, desertification, land degradation, measuring carbon sinks, and drought indicators. The NDVI, and many other AVHRR-derived products from the 30+-year record, were used for a range of conservation applications, including biophysical measures for species distribution modelling, land-cover mapping, and fire management, to name a few (Loveland *et al.* 1995, Sukhinin *et al.* 2004, Buermann *et al.* 2008). Long-term monitoring of biophysical measures and responses to climate oscillations, such as the El Niño Southern Oscillation (ENSO) (Anyamba and Tucker 2005, Pettorelli *et al.* 2005), provided vital information to assess long-term trends and make future projections of environmental change.

Providing continuity for the AVHRR mission was the Moderate Resolution Imaging Spectroradiometer (MODIS) instrument, launched in 1999 on board the Terra satellite, with a duplicate instrument launched on the Aqua satellite in 2002. MODIS was strategically designed to replicate the Landsat sensor design but with a coarser spatial resolution as with as higher temporal resolution, similar to the AVHRR. Multi-disciplinary science teams participated in both the development of the sensor, including sensor band design, as well as the suite of derived products. Many MODIS-derived products have supported a range of conservation applications. These include the early generation of global active fire data (Csiszar *et al.* 2005) to assess the global impacts of fire, the identification of forest-disturbance detection and edge effects (Jin and Sader 2005), the detection of land-cover change based on multi-temporal MODIS imagery (Lunetta *et al.* 2006), and modelling species distribution and richness using the leaf area index (Saatchi *et al.* 2008). Over the sea, it has enabled the measurement of many marine parameters such as sea surface temperature and chlorophyll (Chapter 2). Similarly, ESA invested in the continuity of the SPOT Vegetation programme (operational since 1998) by launching Proba-V in 2013. The sensor on Proba-V is similar to the MODIS sensor.

Most recently, the Landsat Data Continuity Mission (LDCM), renamed Landsat 8 after its 2013 launch, ensures the provision of repeated medium-resolution, multispectral observations. Such observations have proved key to monitoring land cover and land-cover change over the past 40 years. The value of repeated acquisitions has been further amplified by the recent launch of the Sentinel satellite series, as part of ESA's Copernicus programme. Sentinel-2A, launched in June 2015, and Sentinel-2B, launched in March 2017, have Landsat-like sensors with similar spectral and spatial characteristics. These complimentary sensors ensure repeated image acquisitions every 2–3 days, when combined with Landsat. For example, a new Harmonized Landsat-8–Sentinel-2 (HLS) global product is in development and is currently available for selected regions. The development of these products utilises the cloud-computing capabilities of NASA's NEX facility to apply radiometric, spectral, geometric, and spatial adjustments to data from both sensors, to create a combined single product (Claverie *et al.* 2017). The collaborative, long-term vision and coordination among international space agencies to align missions provides a breakthrough solution to the long-standing trade-offs between spatial resolution and temporal resolution.

Monitoring in Real Time

Near-real-time monitoring, afforded by expedited data processing and distribution, is increasingly used by the conservation community. This community discovered the value of near-real-time monitoring in keeping a finger on the pulse of Earth's dynamics in the early 2000s, when the University of Maryland and partners designed a pioneering data-dissemination process, which led to the development of the Fire Information for Resource Management System (FIRMS), an extension of what was originally the MODIS Rapid Response System (Butler 2008, Davies *et al.* 2009). This free, subscription-based system packaged high-level processed data (shapefiles and text files with geolocation points), and emailed these outputs directly to decision-makers. CI, one of the original partners with FIRMS, in turn developed a highly customised system, the Fire Alert System (now called Firecast), designed in collaboration with conservation practitioners in CI's priority geographies (Box 9.3) (Musinsky *et al.* 2018). Specifically, the system gave end-users the freedom to develop customised subscription options, such as being able to select, for example, areas of interest using nationally approved datasets, static map images, language preference, plain text emails, and attachment options. Thus, users could overcome barriers that had previously prevented data use, including geographic information system (GIS) capacity, internet bandwidth, and language barriers. Through Firecast, government agencies and conservation organisations are empowered with vital and time-sensitive information on threats to key areas of importance, such as protected areas (Chapter 5), key biodiversity areas, national forestry zones, environmental investment sites, and agricultural frontier areas. This information enables large-scale monitoring, facilitated immediate decisions to enforce land-use policies, and informed strategic patrolling to conserve resources (Musinsky *et al.* 2018).

BOX 9.3   **CI's Fire Alert System**

Over the years, CI documented success stories of our near-real-time fire alert systems, benefiting conservation actions. One example is the delivery of alerts from CI's Fire Alert System, which led to the arrest of 81 individuals in a single day in 2007. The arrests were prompted by the discovery of illegal clearing, deep inside the Kerinci Seblat National Park in Indonesia, based on a fire detected by satellite observation and sent through an email alert to the park manager, who dispatched rangers the same day (Butler 2008). In another example from Baly Bay in Madagascar, the Durrell Wildlife Conservation Fund engaged communities in conservation by using cash prizes to encourage communities to quickly respond to fire alerts detected inside the habitat of the critically endangered plowshare tortoise (Conservation International 2014). These two examples highlight the diversity of end-users who, when empowered by near real-time monitoring information, can instantly turn the tide of conservation threats to conservation successes.

Another vital monitoring product used by the conservation community in operational systems is vegetation NDVI, for near-real-time detection of deforestation and forest disturbances, alerting decision-makers requiring rapid assessments and response (De Sy *et al.* 2012). This information, on the time-scale of weeks to months, is vital for effectively targeting investments, strategising park ranger patrols, and providing transparent information for policy decisions on land-use management and these systems are operational both globally and regionally. Two global examples are GFW's FORest Monitoring for Action (FORMA) alerts and Quarterly Indicator of Cover Change (QUICC) alerts, produced by NASA Ames Research Center (Hammer *et al.* 2014). One regional example is Datecçãode Desmatamento em Tempo Real (DETER), an early warning system operated by Brazil's space agency that uses MODIS NDVI anomalies to rapidly identify deforestation hotspots, allowing for immediate response by law enforcement. The near-real-time monitoring system, in combination with concerted changes in law-enforcement policies, international funding from REDD+, and social and political campaigns supporting zero-deforestation policies, contributed to a two-thirds reduction in deforestation rates from 2007 through 2014, compared to the previous two decades (Boucher *et al.* 2013). Most recently, these national satellite-based monitoring systems indicate that deforestation rates in Brazil are on the rise again, and thus, have provided transparent evidence that leveraged global attention to Brazil's urgent sustainability predicament.

9.3   INCREASED USE OF REMOTE SENSING DATA IN GLOBAL CONSERVATION

Conservation budgets are limited. This is particularly true in developing countries (McCarthy *et al.* 2012), where conservation is most crucial and policy decisions can result in severe and sometimes irrevocable consequences for biodiversity, livelihoods, and climate change. Already limited conservation budgets rarely cover the purchase of remote sensing imagery, for example. Historically, Landsat was available for purchase with individual images costing up to 800 USD (Wulder *et al.* 2008). Also, the inconsistent archiving of Landsat by different receiving stations around the world resulted in image archive gaps in many areas. In the early 2000s, the Global Land Cover Facility (GLCF), operated by the University of Maryland, began acquiring imagery, including Landsat, to populate its archive and distributed data through the Earth Science Data Interface (ESDI). The GLCF acquired Landsat through multiple avenues, including users donating imagery purchased from the USGS and through direct purchases from NASA, the United Nations, non-governmental organisations, and private companies. Acquired imagery was then made freely available to a wide user community through the ESDI (Saurabh Channan, personal communication, 2017). This process, and the resulting data archive, represented a huge asset to conservation practitioners. With access to free data, they could readily develop baseline maps of habitat extent and condition, generate more frequent mapping of priority conservation landscapes around the world, and perform regular monitoring activities.

The true game changer for conservation applications was the free release of Landsat data in 2008 in a GIS-friendly format (Goetz *et al.* 2015, Song *et al.* 2015, Turner *et al.* 2015, Wulder *et al.* 2016). This USGS/NASA initiative for free and open data meant new global standards for data access and ESA followed suit allowing free access to the Copernicus mission data. All NASA Earth observation data products are now free and open access (www.nasa.gov/sites/default/files/atoms/files/206985_2015_nasa_plan-for-web.pdf c). In addition, the Landsat archive is now also available in a GIS-friendly geotiff format, eliminating the need for specialised and expensive image processing software or customised tools. CI, like many conservation organisations, took full advantage of this release, which enabled scientists all over the world to acquire and analyse data for their conservation applications at no cost. Analysts exploited the entire historical Landsat archive by developing processing streams that, for example, significantly reduced the impact of clouds (persistent cloud cover in tropical regions has historically impacted conservation mapping activities). The technical barriers faced by conservation practitioners were reduced by the provision of cloud-masked and atmospherically corrected images, for example. This is particularly valuable for those working in developing countries, where limited capacity, in terms of both software and technical proficiency, have historically represented challenges for using these image sources. Access to free, medium-spatial-resolution optical data has opened the door to operational forest monitoring and enabled the development, and routine update, of global- to national-level forest cover and change products. These, in turn, have supported a host of conservation-related initiatives, such as providing annual forest monitoring for countries.

The free availability of radar data will open more doors. While missions such as ALOS and ALOS-2, collaborative endeavours between NASA and the Japan Aerospace Exploration

Agency (JAXA), acquire Phased Array type L-band Synthetic Aperture Radar (PALSAR) imagery, these data were only released at no cost as annual composites, with a spatial resolution of 50-m. The Sentinel-1A and -1B sensors, part of the ESA's Copernicus programme, are of particular interest to the conservation community because these data are available at no cost and, as with all radar sensors, they can penetrate clouds (Chapter 2). Additionally, there is a potential for these instruments to measure canopy structure, providing additional land-cover information to complement optical sensors. While there is a learning curve involved in using these data, both in terms of pre-processing as well as interpretation and understanding the physical properties of radar data, ESA provides free toolboxes, such as the Sentinel Application Platform (SNAP), to process the data and help inform the user community. Yet, even with free data and tools, barriers remain to widespread use, including the sheer volume of the data, which require ample internet bandwidth, storage capacity, and computing capacity.

The rapid evolution of open-source tools and software presents another key advancement, amplifying the uptake of remote sensing data and derived products by the conservation community. Previously, proprietary, often expensive, software restricted the use of remote sensing and GIS data by many conservation practitioners, especially in developing countries. The explosion of open-source tools and visualisation platforms, such as Google Earth (www.google.com/earth – originally Keyhole EarthViewer), GRASS (https://grass.osgeo.org/), Q-GIS (www .qgis.org), R (www.r-project.org), as well as GEE (https://earthengine.g oogle.com/), to name a few, reflects this revolution. In fact, the visualisation capabilities, remote sensing image archives and computing power provided by GEE and similar platforms has facilitated time-series analyses, using entire archives of remote sensing imagery as well as accuracy assessment estimations. GEE and other platforms and web tools are rapidly facilitating data interpretation, dissemination, pre-processing, analysis, and accuracy assessment for CI's applications.

9.4   RECENT REMOTE SENSING CONSERVATION
CHALLENGES

Even with rapid advances in technology and remote sensing conserva-tion applications, conservation practitioners are still challenged by limited budgets and technical capacity. The advancements of the past decade have set the bar higher, to generate more complex products at regional to global scales, which inform policies aimed at achieving ambitious national and global conservation and sustainability targets. For example, challenges in accurate ecosystem mapping to inform sustainable land-management decisions (i.e. mapping natural capital, mapping blue carbon, evaluating ecosystem health, generating accurate biomass and emission estimates to inform policies) can be addressed by recent advances in remote sensing, such as the increasing prevalence of very high-spatial-resolution (VHR) optical data and non-optical sensors like lidar and high-resolution radar. However, conservation practi-tioners have yet to fully adopt these types of observations. Accessibility is a problem. Cost presents a barrier, due to the high acquisition costs of VHR optical and lidar data and, until recently, radar data. Data of <10-m spatial resolution are only available commer-cially. Further, the limited footprint of these VHR data results in limited spatial coverage and low temporal frequency. Radar sensors capture information differently to optical sensors and this requires a different technical skillset to interpret and analyse the imagery, a skillset not prevalent in the conservation remote sensing community. The lidar data, while highly desired by conservation practitioners for measures of carbon stock and under-canopy ecosystem disturbances, remains out of reach to many due to the high costs of acquiring the data and the technical expertise required to process and analyse the multi-dimensional aspect of these data. Further, use of these data frequently necessitates advanced image-processing techniques, requiring proprie-tary software, and significant storage, capacity, and computing require-ments. This is further confounded by internet bandwidth limitations, which present a continual barrier in many regions.

9.5 RECOMMENDATIONS

Of upmost importance to the continuation of uptake of technological advances by the conservation community is the continued trend for free, higher-level data products and open-source tools that are easily accessible. By accessible, we mean that they should be easy to find online, and also useable, regardless of situational internet bandwidth and computing performance. GEE is an example of this, and could represent a major step forward for conservation practitioners by overcoming previous barriers to data access and analysis. GEE hosts the entire Landsat archive, as well as many other satellite remote sensing products, such as MODIS and AVHRR NDVI. In addition, GEE is currently incorporating Sentinel data and providing pre-processed radar data products. Users with internet connectivity and modest Java-script coding abilities can visualise and analyse these data in a web browser. GEE has a rapidly growing user base, who utilise GEE for wide-ranging applications and who share models based on open-source code. CI is currently incorporating GEE into several of its conservation activities. For example, GEE is the basis for a set of tools under development as part of a Global Environmental Facility-funded project to monitor land degradation and to assess the impacts of rangeland management interventions. GEE is also being used at CI to streamline mapping, monitoring, and reporting of forest change and carbon-emission efforts, and the associated development of institution-wide tools, as well as capacity-building activities to promote the use of remote sensing for systematic monitoring.

Accessibility alone does not lead to widespread use of data. A concerted effort in capacity-building is required within the conservation community, and beyond, to enable the optimal use of existing data and tools. The goal should be to build permanent global capacity for remote sensing analysis. Many initiatives are addressing this issue. Three such initiatives are NASA's Applied Remote SEnsing Training (ARSET) programme and SilvaCarbon and the US Forest Services international programme. NASA's ARSET programme provides capacity-building for a range of remote sensing applications, providing free webinars to remote sensing analysts, including conservation practitioners, around the globe, on various aspects of remote sensing from theory to specific applications (e. g. fire management, national forest monitoring systems, air-quality monitoring). SilvaCarbon is a US technical cooperation initiative, combining the technical capacity of multiple federal agencies as well as US universities and non-governmental organisations to provide capacity-building to selected tropical countries that are developing monitoring systems in support of initiatives such as REDD+. One of the major activities of SilvaCarbon is to provide capacity-building in a host of remote sensing techniques to countries who are developing their NFMSs. The US Forest Service's international programme, together with non-governmental organisations and academic institutions, promotes international capacity-building through in-person trainings, workshops, and joint projects.

From working in international capacity-building for more than 15 years, one of the biggest challenges that we here at CI observe is building permanent capacity in-country. This is not unique to CI, but is widespread across the conservation community (de Klerk and Buchanan 2017; Palumbo *et al.* 2017). The permanence of in-country technicians in key roles presents a continual problem, as turnover in many governments and in-country non-governmental organisations can be high. To ensure a level of redundancy that negates the impacts of high turnover, we must train more technicians and, further, promote ethnic and gender diversity for both trainers and trainees (Clark *et al.* 2016). SilvaCarbon and the Society for Conservation GIS have programmes to train technicians in-country. The SERVIR–Mekong programme is another excellent example, promoting gender equality in technical capacity-building trainings (SERVIR–Mekong 2015). Further, the conservation remote sensing community should implement iterative processes to evaluate immediate and longer-term learning retention, following trainings, in the same way that we monitor the efficacy of conservation interventions. Again, the ARSET programme is a good example of this effort. Every training activity is followed by a participant survey to capture how many students apply the knowledge they learn from the webinar series. These types of training events highlight a smart, efficient method for amplifying efforts in capacity-building.

9.6   THE FUTURE ROLE OF REMOTE SENSING IN CONSERVATION

We believe the future of conservation remote sensing will combine holistic ecosystem monitoring and forecasting for natural, production, and urban systems, including mixed landscapes. This will enable further operational monitoring and near-real-time analyses of ecosystem function, as well as assessments of broader ecosystem traits, including ecosystem health and human welfare. To date, technologies and applications have seen monitoring tools and products develop from single-point static maps to dynamic analyses. Often, these are delivered in real time and allow for early warning and near-term forecasting. One current example is CI's Firecast system, which allows on-demand analyses and visualisations of landscape threats. This is updated in near real time, including daily fire weather based on satellite observations of weather conditions (Steininger *et al.* 2010), active fire alerts from MODIS and the Visible Infrared Imaging Radiometer Suite (VIIRS) (Giglio *et al.* 2003, Schroeder *et al.* 2014, Chapter 5), QUICC forest disturbance alerts (Potter *et al.* 2005), and fire season severity forecasts for the Amazon (Chen *et al.* 2011). In one pilot site, the Alto Mayo Forest Reserve in Peru, Firecast integrates an *in situ* acoustic network, calibrated to detect chain-saw sounds, which alerts park managers to illegal logging operations. Firecast is only one example of the recent trend to integrate satellite data with field-based sensors, crowd-sourced data from mobile devices, and data from unmanned aircraft vehicles (UAVs) and other airborne platforms.

Holistic monitoring also requires the ready provision of advanced mapping and monitoring products. To date, the community has achieved significant advances in the generation of global products using Earth-observation data. These include operational forest cover and change mapping (Hansen *et al.* 2013, Kim *et al.* 2014), global-scale mangrove mapping (Giri *et al.* 2011), and global-scale land cover and change (e.g. ESA's 300-m products for 1992–2015, generated through the ESA Climate Change Initiative–Land Cover project). These impressive advanced products were almost unthinkable to develop less than a decade ago, and their availability provides the backbone of many applications.

Immediate challenges are the development of readily accessible approaches for the mapping and monitoring of forest degradation, as well as approaches to facilitate accounting for greenhouse gas inventories, natural capital accounting efforts, and the United Nations Sustainable Development Goal (SDG) indicators. Improving our understanding of natural ecosystems beyond simply ecosystem extent is critical to performing assessments of ecosystem health and to valuing ecosystems for the critical services they provide for human well-being. Monitoring multiple changes on Earth, including climate, air quality, urban landscapes, mixed agricultural landscapes, and commercial supply-chain sourcing, informs the pulse of the Earth's biosphere. This further enables us to identify alternative development paths and to evaluate the impact of policies that are intended to achieve conservation goals.

Achieving seamless data integration and advances in environmental monitoring and forecasting will be possible if we continue investments in cloud-based platforms designed to increase access to higher-level data products, and intuitive tools for data visualisation and analysis. Diverse organisations, including technology companies, software companies, government agencies, academic institutions, and non-profit organisations are collaborating on the development of online platforms for data visualisation and analysis, data aggregation and dissemination, and research on new methods for data analysis. Private-sector companies are ahead of the curve, having developed expensive integrated monitoring systems for large-scale agriculture management and monitoring commodity sourcing. One example is the Starling satellite service by Airbus, developed in partnership with the Forest Trust and SarVision. This sophisticated agriculture monitoring system combines radar and high-resolution optical imagery to identify crop types and attribute deforestation to help large corporations achieve their net-zero deforestation commitments. While conservation is unlikely to have the same resourcing as the private sector, this example illustrates the potential of remote sensing if the community could garner additional financial resources and leverage the innovations that have been developed for commercial purposes.

Democratising these valuable data and tools, coupled with capacity-building, can empower a host of stakeholders with key information that is needed to manage their land and natural resources. Small actions from well-informed decisions can amplify to global significance. Tools to support protected-area managers, for example, contribute toward the development and success of a country's NFMS, and thus toward meeting national emission reduction targets. Collectively, countries that improve their forest monitoring systems are contributing to global climate mitigation goals. Other stakeholders beyond government agencies may also have large impacts on reaching global sustainability goals. Examples include local communities and indigenous peoples. Empowering such populations with Earth-observation data and tools can potentially improve land management and protect their territories from illegal development. One such initiative is the World Bank's Dedicated Grant Mechanism for Indigenous Peoples. In another initiative, Google, through its Google for Good programme, has equipped indigenous communities in Brazil with Earth observation capabilities to perform monitoring of their indigenous lands.

On the immediate horizon for conservation remote sensing are upcoming satellite missions with reliable and repeated observations that inform measurements of water, biomass, and canopy structure. These missions are anticipated to fill critical gaps in data that have been identified by the conservation community. We look forward to the entire implementation of the ESA's Copernicus programme and, particularly, the range of applications that will be afforded by the Sentinel satellites. The complementarity of the Landsat and Sentinel 2 instruments highlights the effectiveness of inter-agency and international coordination through aligned missions to overcome technology limitations. In this case, we see increased revisit rates without compromising spatial resolution. The conservation community is encouraged about plans for a Landsat 9 mission, as Landsat continuity is core to conservation applications. Additionally, new data sources of particular interest to the conservation community include: (i) the NASA–Indian Space Research Organisation (ISRO) synthetic aperture radar (NI-SAR) L-band and S-band radar, designed to monitor ground water, biomass, and ecosystem disturbances (anticipated launch 2020); (ii) Global Ecosystem Dynamics Investigation (GEDI; an instrument that will be on the International Space Station) lidar for measuring biomass, canopy structure, and height (anticipated launch 2019); and (iii) a new P-band radar from ESA's Earth Explorer Biomass mission, measuring forest canopy height and biomass (anticipated launch 2021). Data from these missions will help with priority-setting for investments that maximise conservation of critical ecosystem services, monitor landscape degradation and deforestation, and assess ecosystem function for valuing natural capital.

Private satellite companies who have entered the remote sensing sphere will undoubtedly play a larger role in conservation in the future. For example, there is much anticipation over the potential applications of nanosatellites, also called cubesats, for conservation applications. Nanosatellites are low-cost sensors, designed to work through a connected network of sensors, for continuous, high-spatial-resolution monitoring. Nanosatellites are not necessarily designed specifically for natural-resource management and thus their potential must be harnessed creatively. While the spectral resolution is not ideal for measuring or mapping landscape characteristics, the continuous monitoring of the Earth's surface is extremely valuable for near real-time applications, through early warning systems alerting to changes in the landscape and monitoring natural disasters. These systems can potentially achieve wide-ranging conservation successes through effective partnerships with conservation practitioners.

9.7   CONCLUSION

In the past two decades, the conservation community has capitalised on a suite of advances in remote sensing technologies to help address conservation challenges. Advances in sensors that are designed specifically for monitoring landscapes, together with continuity and inter-agency alignment of missions, repeated and expedited monitoring information and alert tools, and investment in readily accessible data have all combined together with an explosion of open-source tools to make the use of remote sensing technology in conservation applications compulsory. Remote sensing is now routinely relied upon for a range of conservation applications, such as prioritising conservation investments, informing national and international policy decisions, incorporating natural assets into national accounting, and facilitating rapid responses to emerging ecosystem threats. These advances, addressing the conservation community's needs, are attributable to multiple actors, including satellite sensor engineers, the research community, the applications community, the private sector, and champions of the open-source movement. The acceleration in technological development in recent years has expanded the realm of possibility for conservation remote sensing applications. Continued support for elements of satellite design, routine monitoring, mission continuity, and accessibility through open-source data and tools are essential for advancing conservation initiatives. An additional emphasis to integrate high-resolution and active sensors (e.g. lidar and radar) as well as building global capacity are critical for the uptake of these valuable yet under-utilised technologies. The pathway to global well-being hinges on integration, collaboration, and democratisation of remotely sensed data and tools for scaling-up conservation successes, from site to global scales.

REFERENCES

Angelsen, A., Brown, S., Loisel, C., *et al.* (2009). Reducing Emissions from Deforestation and Forest Degradation (REDD): An Options Assessment Report. Meridian Institute. See www.redd-oar.org/links/REDD-OAR_en.pdf.

Anyamba, A. and Tucker, C. J. (2005). Analysis of Sahelian vegetation dynamics using NOAA–AVHRR NDVI data from 1981–2003. *Journal of Arid Environments*, **63**, 596–614.

Boucher, D., Roquemore, S., and Fitzhugh, E. (2013). Brazil's success in reducing deforestation. *Tropical Conservation Science*, **6**, 426–445.

Buermann, W., Saatchi, S., Smith, T., *et al.* (2008). Predicting species distributions across the Amazonian and Andean regions using remote sensing data. *Journal of Biogeography*, **35**, 1160–1176.

Butler, R. A. (2008). Fire monitoring by satellite becomes key conservation tool: an interview with GIS experts at Conservation International and the University of Maryland. Conservation International. See https://news.mongabay.com/2008/03/fire-monitoring-by-satellite-becomes-key-conservation-tool/.

Chen, Y., Randerson, J. T., Morton, D. C., *et al.* (2011). Forecasting fire season severity in South America using sea surface temperature anomalies. *Science*, **334**, 787–791.

Clark, B. L., Bevanda, M., Aspillaga, E., and Jørgensen, N. H. (2016). Bridging disciplines with training in remote sensing for animal movement: an attendee perspective. *Remote Sensing in Ecology and Conservation*, **3**, 30–37.

Claverie, M., Masek, J.G., Junchang, J., and Dungan, J. L. (2017). Harmonized Landsat-8 Sentinel-2 (HLS) product user's guide. Harmonized Landsat–Sentinel-2 (HLS) project. NASA. See https://hls.gsfc.nasa.gov.

Conservation International (2014). Conservation tools: satellites sound fire alarm in tropical forests. See http://blog.conservation.org/2014/07/conservation-tools-satellites-sound-fire-alarm-in-tropical-forests/.

Csiszar, I., Denis, L., Giglio, L., Justice, C. O., and Hewson, J. (2005). Global fire activity from two years of MODIS data. *International Journal of Wildland Fire*, **14**, 117–130.

Davies, D. K., Ilavajhala, S., Minnie Wong, M., and Justice, C. O. (2009). Fire information for resource management system: archiving and distributing MODIS active fire data. *IEEE Transactions on Geoscience and Remote Sensing*, **47**, 72–79.

De Sy, V., Herold, M., Achard, F., *et al.* (2012). Synergies of multiple remote sensing data sources for REDD+ monitoring. *Current Opinion in Environmental Sustainability*, **4**, 696–706.

De Wulf, R. R., Goossens, R. E., MacKinnon, J. R., and Cai, W. S. (1988). Remote sensing for wildlife management: giant panda habitat mapping from Landsat MSS images. *Geocarto International*, **3**, 41–50.

Gaveau, D., Epting, J., Lyne, O., *et al.* (2009). Evaluating whether protected areas reduce tropical deforestation in Sumatra. *Journal of Biogeography*, **36**, 2165–2175.

Giglio, L, Descloitres, J., Justice, C. O., and Kaufman, Y. (2003). An enhanced contextual fire detection algorithm for MODIS. *Remote Sensing of Environment*, **87**, 273–282.

Giri, C., Ochieng, E., Tiezen, L. L., *et al.* (2011). Status and distribution of mangrove forests of the world using Earth observation satellite data. *Global Ecology and Biogeography*, **20**, 154–159.

Goetz, S. J., Hansen, M., Houghton, R. A., *et al.* (2015). Measurement and monitoring needs, capabilities and potential for addressing reduced emissions from deforestation and forest degradation under REDD+. *Environmental Research Letters*, **10**, doi: 10.1088/1748–9326/10/12/123001.

Hammer, D., Kraft, R., and Wheeler, D. (2014). Alerts of forest disturbance from MODIS imagery. *International Journal of Applied Earth Observation and Geoinformation*, **33**, 1–9.

Hansen, M. C., Potapov, P. V., Moore, R., *et al.* (2013). High-resolution global maps of 21st-century forest cover change. *Science*, **342**, 850–853.

Harper, G. J., Steininger, M. K., Tucker, C. J., Juhn, D., and Hawkins, F. (2007). Fifty years of deforestation and forest fragmentation in Madagascar. *Environmental Conservation*, **34**, 325–333.

Jin, S. and Sader, S. A. (2005). MODIS time-series imagery for forest disturbance detection and quantification of patch size effects. *Remote Sensing of Environment*, **99**, 462–470.

Kim, D-H., Sexton, J. O., Noojipady, P., *et al.* (2014). Global, Landsat-based forest-cover change from 1990 to 2000. *Remote Sensing of Environment*, **155**, 178–193.

De Klerk, H. M. and Buchanan, G. (2017). Remote sensing training in African conservation. *Remote Sensing in Ecology and Conservation*, **3**, 7–20.

Loveland, T. R., Merchant, J. W., Brown, J. F., *et al.* (1995). Seasonal land-cover regions of the United States. *Annals of the Association of American Geographers*, **85**, 339–355.

Luque, S. S. (2000). Evaluating temporal changes using Multi-Spectral Scanner and Thematic Mapper data on the landscape of a natural reserve: the New Jersey Pine Barrens, a case study. *International Journal of Remote Sensing*, **21**, 2589–2610.

Lunetta, R. L., Knight, F. K., Ediriwickrema, J., Lyon, J. G., and Worthy, L. D. (2006). Landcover change detection using multi-temporal MODIS NDVI data. *Remote Sensing of Environment*, **105**, 142–54.

McCarthy, D. P., Donald, P. F., Scharlemann, J. P., *et al.* (2012). Financial costs of meeting global biodiversity conservation targets: current spending and unmet needs. *Science*, **338**, 946–949.

Palumbo, I., Rose, R. A., Headley, R. M. K., *et al.* (2017). Building capacity in remote sensing for conservation: present and future challenges. *Remote Sensing in Ecology and Conservation*, **3**, 21–29.

Pettorelli, N., Vik, J. O., Mysterud, A., *et al.* (2005). Using the satellite-derived NDVI to assess ecological responses to environmental change. *Trends in Ecology & Evolution*, **20**, 503–510.

Portela, R., Nunes, P. A. L. D., Onofri, L., *et al.* (2012). Assessing and valuing ecosystem services in Ankeniheny–Zahamena Corridor (CAZ), Madagascar: a demonstration case study for the Wealth Accounting and the Valuation of Ecosystem Services (WAVES) Global Partnership. Conservation International. See www.wavespartnership.org/sites/waves/files/images/WAVES_Madagascar_Report.pdf.

Saatchi, S., Buermann, W., Ter Steege, H., Mori, S., and Smith, T. B. (2008). Modeling distribution of Amazonian tree species and diversity using remote sensing measurements. *Remote Sensing of Environment*, **112**, 2000–2017.

Schroeder, W., Oliva, P., Giglio, L., and Csiszar, I. A. (2014). The new VIIRS 375 m active fire detection data product: algorithm description and initial assessment. *Remote Sensing of Environment*, **143**, 85–96.

SERVIR–Mekong. (2015). Gender and GIS: guidance notes. Bangkok: Asian Disaster Preparedness Center. See https://servir.adpc.net/sites/default/files/public/publications/attachments/Gender-GIS-2015.pdf.

Song, X. P., Huang, C., Saatchi, S. S., Hansen, M. C., and Townshend, J. R. (2015). Annual carbon emissions from deforestation in the Amazon Basin between 2000 and 2010. *PLOS ONE*, **10**, e0126754.

Steininger, M. K., Tucker, C. J., Townshend, J. R. G., *et al.* (2001). Tropical deforestation in the Bolivian Amazon. *Environmental Conservation*, **28**, 127–134.

Steininger, M. K., Tabor, K., Small, J., *et al.* (2013). A satellite model of forest flammability. *Environmental Management*, **52**, 136–150.

Sukhinin, A. I., French, N. H. F., Kasischke, E. S., *et al.* (2004). AVHRR-based mapping of fires in Russia: new products for fire management and carbon cycle studies. *Remote Sensing of Environment*, **93**, 546–564.

Tabor, K., Burgess, N., Mbilinyi, B., Kashaigili, J., and Steininger, M. K. (2010). Forest and woodland cover and change in coastal Tanzania and Kenya, circa 1990 to circa 2000. *The Journal of East African Natural History*, **99**, 19–45.

Tucker, C. J. (1979). Red and photographic infrared linear combinations for monitoring vegetation. *Remote Sensing of Environment*, **8**, 127–150.

Turner, W., Rondinini, C., Pettorelli, N., *et al.* (2015). Free and open-access satellite data are key to biodiversity conservation. *Biological Conservation*, **182**, 173–176.

Wulder, M. A., White, J. C., Goward, S. N., *et al.* (2008). Landsat continuity: issues and opportunities for land cover monitoring. *Remote Sensing of Environment*, **112**, 955–969.

Wulder, M. A., White, J. C., Loveland, T. R., *et al.* (2016). The global Landsat archive: status, consolidation, and direction. *Remote Sensing of Environment*, **185**, 271–83.

10 Operational Conservation Remote Sensing
Common Themes, Lessons Learned, and Future Prospects

Allison K. Leidner and Graeme M. Buchanan

10.1 INTRODUCTION

Conservation is complex, making it no different from many other major challenges faced by humanity. The urgency of the biodiversity crisis means conservationists cannot baulk at this complexity, and instead need to move quickly to action. The case studies in the preceding chapters illustrate an informed, 'can do' attitude. They demonstrate how remote sensing helps to solve pressing conservation problems and they focus on the application of remote sensing to clear local or regional conservation needs. This user-driven approach is markedly different from more academic demonstration projects where applications are presented to users in the hope that they will find a problem to which it can be applied. Each case study describes how the actions of a collaborating team of scientists and conservation managers resulted in the use of scientific advances and satellite remote sensing for conservation. The case studies describe how the teams leveraged the value of satellite remote sensing for their objectives, and then established pathways to incorporate Earth observations into conservation activities in an operational, ongoing way. This latter point is key for 'mainstreaming' remote sensing into conservation, by which we mean making remote sensing a standard part of the conservationists' toolbox.

In this final chapter, we draw attention to common themes that have emerged from the preceding case-study chapters. For each study, we originally asked the authors not only to outline the conservation issue and scientific methods employed to address the problem but also to discuss how the collaborations came about, the challenges faced, and the lessons learned. This narrative aspect is often absent from scientific journal articles as they understandably focus on completely objective statements that can be quantified and referenced. In developing this book, we sought an approach that would complement the scientific focus of journal articles, and we encouraged contributors to present each case study as a full story in a discussion format that focused on intertwining the scientific and human side of each project. We believe that each of the chapters has delivered on this objective, albeit after some encouragement by the editors to pay more attention to the mistakes and missteps. As a conservation community, we need to more openly discuss what does not work to advance conservation, and why, in order to make the overall endeavour more successful (Redford and Taber 2000).

As noted in Chapter 1, we sought out specific case studies where satellite remote sensing was being used in an operational context. This meant we did not include some well-regarded projects that have utilised satellite remote sensing in a one-off way, for example to set priorities. Additional limitations meant we could not highlight other great examples, such as Coral Reef Watch (https://coralreef watch.noaa.gov) and the Global Surface Water product (Pekel *et al.* 2016; https://global-surface-water.appspot.com/). Some project leaders we contacted were understandably too busy to write up a case study on our schedule and we are very grateful to those that could take the time to contribute to this book. As editors, we were also mindful of the disproportionate attention given to conservation projects in Europe, North America, and Australia (Lawler *et al.* 2006, Burgman *et al.* 2015). We did succeed in finding case studies from outside these regions – half the projects in this book are located in Africa and Asia. However, the authorship does not reflect this geographic distribution, as all but one author is currently working at an institution in the United States, Australia, or Europe. Nonetheless, we strongly believe the case studies in the preceding chapters provide excellent insight into conservation remote sensing applications.

10.2   COMMON THEMES, LESSONS LEARNED, AND IMPLICATIONS

The six case studies are examples of aquatic and terrestrial ecosystems from around the world. Each conservation problem and solution had individual political, cultural, and ecological circumstances. Below, we highlight common themes and lessons learned that emerged, and discuss their implications.

Collaboration

Each case study represents successful formal or informal collaborations of scientists and conservation managers. Most chapters have authors from both communities and even those that were authored by one individual were bolstered by partnerships. This cooperation is exactly what several conservation community workshops have called for when assessing how to better utilise remote sensing for conservation (see Chapter 1) as there can be a gap between what is thought desirable and what is actually feasible when addressing a conservation challenge, but collaboration proved to be key. Interestingly, the conservation side led the charge in all of the case studies, and there was significant end-user engagement and involvement from the inception of the project. This underscores the notion that a key to successful conservation action is to address a clear problem, and conservationists are often those most informed about the key pressures on biodiversity. Many of the chapter authors who had a strong technical background in remote sensing also came from a strong conservation background. This highlights the benefits of better integrating remote sensing into conservation-related training at universities and elsewhere, and supports making additional investments in capacity-building (see the section on building and enabling remote sensing capacity later in this chapter for further discussion). However, there may still be gaps in knowledge, as professionals understandably specialise in a given field. Conservation experts with a foot in the remote sensing camp might not be up to date on the latest advances in remote sensing science.

Likewise, remote sensing scientists with years of technical training are likely to be proficient in the latest advances in their field, but unfamiliar with the detailed ways in which conservationists identify challenges, derive response, and execute solutions. Limited connectivity between these groups may be one reason why operational conservation projects do not use observations from the newer technologies of lidar and radar missions more frequently.

Free, Open, and Accessible Data

Freely available and accessible satellite observations have had a very notable impact in conservation. Every case study used free observational products from US government agencies (the National Aeronautics and Space Administration (NASA), the US Geological Survey (USGS), and the National Oceanic and Atmospheric Administration (NOAA)) and/or the European Space Agency (ESA) and the European Commission (EC). Nearly all of the projects noted that the availability of data were key to their project's success. Tabor and Hewson (Chapter 9) were particularly strong in their emphasis of this point so we do not go into additional depth here. Ultimately, good information is the cornerstone of good decision-making, but that can only be realised if the information is available as widely as possible, and thus available for creative and urgent applications.

Space agencies are interested in increasing the user base for their satellite observations (Paganini *et al.* 2016), and we encourage the conservation community to continue to promote the importance of free and open data policies. Space agencies have long served the physical sciences and biogeochemistry communities, but are relatively unfamiliar with the needs of the biodiversity research and conservation application communities. Conservationists can provide further encouragement for free and open data policies by raising awareness within space agencies of comparatively new applications and by highlighting how many organisations already use this information. They can also encourage these and other organisations to develop higher-level products that are tailored for conservation needs. In the conservation community, such efforts will not only help local, regional, and national conservation efforts, but will also help to meet the needs of policy assessments (e.g. the Intergovernmental Science-Policy Platform on Biodiversity and Ecosystem Services, better known as IPBES) and international policy agreements (e.g. the Aichi targets under the Convention on Biological Diversity), which numerous nations have agreed to support.

Access to data will also be improved if users need to do less pre-processing, as is the case when observations are available as higher-level data products (see Chapter 2). The Committee on Earth Observation Satellites (CEOS), the international coordinating body for civilian space agencies, has consequently placed a focus on analysis ready data (ARD). ARD centres on the production of satellite remote sensing products that are 'processed to a minimum set of requirements and organised into a form that allows immediate analysis with a minimum of additional user effort and interoperability both through time and with other datasets' (CEOS 2017). Recently, the USGS announced the release of Landsat ARD for the USA; work is underway to finalise the global dataset (https://landsat.usgs.gov/ard). Top-of-atmosphere reflectance, brightness temperature, surface reflectance, and pixel quality assessment will be available for every Landsat tile. The case studies all make use of satellite data products that have been radiometrically and geometrically corrected, and nearly all take advantage of data processed to even higher levels. The new Landsat data will further reduce barriers to use. However, even in cases where higher-level products are available and accessible, conservation activities will still likely need to further tailor such products to their needs.

Spatial and Temporal Resolution of Satellite Observations

Although the conservation community has historically called for very high-spatial-resolution data (e.g. Green *et al.* 2011), we did not see this request heavily emphasised in the case studies presented here. All of the chapters utilised two main sources of freely available satellite data, though many leveraged the use of multiple observing platforms. Murray (Chapter 3) and Jantz *et al.* (Chapter 4) used NASA/USGS Landsat data. These are available at a 30-m spatial resolution with a maximum temporal resolution of 16 days. Palumbo *et al.* (Chapter 5), Escribano and Fernández (Chapter 6), Hebblewhite *et al.* (Chapter 7), and Bailey *et al.* (Chapter 8) used data from Moderate Resolution Imaging Spectroradiometer (MODIS) instruments on NASA's Terra and Aqua satellites. Despite their medium to low spatial resolution (250–1,000 m), the MODIS instruments view Earth twice a day, which was cited as a deciding factor for many groups to utilise MODIS over Landsat data.

An emphasis on the benefits of higher temporal versus spatial resolution for conservation efforts (i.e. MODIS versus the Landsat sensors) is a general trend that we, the editors, have more recently noted. Higher temporal resolution is valuable for the detection of habitat changes over short time periods. Detecting these events rapidly can alert natural-resource managers to environmental change (e.g. fires in Chapters 5 and 9). Higher temporal resolution is also needed to develop information on intra-annual changes in land cover, which can identify anomalies when conditions deviate from the norm (e.g. seasonality in Chapter 6). Finally, higher temporal resolution means that the chance of capturing a cloud-free image of an area is increased, given that passive optical satellites cannot penetrate clouds. As highlighted in Chapter 9, this is particularly valuable for biodiversity-rich tropical regions.

The two case studies that relied on Landsat data focused on changes in habitat extent for species of conservation concern, although they had different reasons for using this satellite system. Murray (Chapter 3) benefited from the higher spatial resolution because tidal flats often have a narrow spatial configuration. However, his primary reason for using Landsat data was the longer observational record, which allowed him to measure the extent of habitat loss, a variable used in the conservation assessment of migratory shorebirds. Spatial resolution was the driving factor for Jantz *et al.* (Chapter 4), as Landsat allowed for a better mapping of chimpanzee habitat. Furthermore, this project was the only one to make use of commercial very high-spatial-resolution (≤1 m) observations, which they received free under an agreement between DigitalGlobe and the Jane Goodall Institute.

For certain applications, such as assessing the condition of rangelands (Chapter 7) and ocean (Chapter 8), higher-spatial-resolution data may not be particularly helpful, as the areas analysed can be extensive. Additionally, limits in computing power would inhibit many groups from processing datasets that had both a high spatial and temporal resolution. Indeed, data-processing issues were noted in both chapters that relied on Landsat data. In the near future, cloud-computing resources, such as those utilised by Jantz *et al.* (Chapter 4) and discussed in Tabor and Hewson (Chapter 9), may render this concern obsolete.

We suggest that it could be instructive to develop quantitative assessments of the absolute and relative trade-offs in having higher-spatial- or higher-temporal-resolution observation for conservation needs. The findings could inform priorities for planned or prospective satellite observing systems and constellations (groups of satellites working together). Such an analysis may also be particularly insightful when new satellite systems are launched that are very similar to existing missions. An example of a collaborative activity is the ongoing work to harmonise the observations of Sentinel 2 and Landsat, resulting in products that have a 30-m spatial resolution, but a temporal resolution of 2 to 3 days (Claverie *et al.* 2017).

Types of Satellite Observations Used

All six case studies relied on passive optical data, which are observations made by satellites that measure energy reflected from the Sun or emitted by the Earth (see Chapter 2). These include Landsat satellite sensors, the AVHRR instruments, the MODIS instruments, and now the Sentinel-2A and Sentinel-2B satellite sensors. The reliance on these observations is understandable. First, the decades-long records of Landsat and AVHRR and the nearly two-decade record of MODIS means that conservationists can take a backward look at a region and make decisions about current situations in the light of present conditions (e.g. Chapters 3, 5, and 6). Access to this back catalogue of data is clearly valuable. It justifies the efforts made by NASA and the USGS in data product development and distribution for Landsat and MODIS data prgrammes. There are more higher-level products derived from MODIS data than from Landsat data, despite Landsat sensors having been in operation for longer than MODIS. These higher-level data products make it easier to conduct analyses across scenes. Landsat is likely to catch up in this regard, with NASA invest-ments in data delivery (e.g. https://landsat.usgs.gov/web-enabled-landsat-data-weld-projects) that have made and will make it easier to analyse Landsat observations across years and scenes. The collation of many of these data in the Google Earth Engine also facilitates that rapid comparison of long-term datasets.

Continuous records and ease of use also play into the second point, which is that a longer observational history likely means that the conservation community is more familiar with how to analyse and interpret these observations for their needs. Over time, a familiarity with passive data, combined with capacity-building efforts, has built strong capabilities to exploit these observations. Furthermore, passive observations are intuitive for many of those unfamiliar with remote sensing, as the sensors work in ways that are analogous to human eyes. We have found that using this comparison is an extremely quick and effective way to introduce almost anyone to remote sensing (A. Leidner, personal observation). The intuitive nature facilitates explanation, and hopefully buy-in, from conservation practitioners who are working with remote sensing data. Ultimately, the familiar-ity, availability, and relative ease of use of passive observations facil-itate a ready use by the conservation community.

A corollary to the theme of heavy reliance on passive optical data is the lack of use of observations from active remote sensing instruments, most notably radar and lidar. It is unfortunate that these observations are not yet better utilised, as they can provide unique information on the vertical structure of ecosystems. We infer that many of the same reasons supporting the use of passive optical data pertain to the paucity of use for active data: radar and lidar data have a shorter temporal record, require different (and some would argue more advanced) skills, and are less intuitive compared to passive optical data. Previous investments in radar and lidar satellite and airborne measurements, combined with the growing availability of these types of observations with recent and upcoming space agency launches (e.g. ESA's Sentinel-1 radar measurements and NASA's Global Ecosystem Dynamics Investigation (GEDI) lidar measurements) will expand conservation remote sensing opportunities.

Building and Enabling Remote Sensing Capacity

The value of capacity-building activities was frequently highlighted, especially so in Tabor and Hewson (Chapter 9). Exposure to satellite remote sensing, familiarity with the technical concepts, and awareness of successful applications should encourage more conservationists to use satellite remote sensing. Each of the case studies successfully engaged skilled remote sensing scientists, leading to the technical incorporation of remote sensing into the conservation application. This illustrates the benefits to projects of either having conservationists with remote sensing skills or a close pairing of remote sensing scientists with conservationists. From our perspective of the six case studies, and our general experiences, both pathways of improving remote sensing capacity are equally constructive.

Developing training materials is time-consuming and expensive, and attention to these activities tends to come at the end of projects, when timelines and budgets are heavily stretched. Thus, even projects that specifically intend to invest in capacity-building efforts sometimes fall short. Consequently, initiatives focused on building remote sensing capacity more generally, such as the NASA Applied Remote SEnsing Training (ARSET) programme (discussed in Chapter 2), provides an important way to improve capacity. In Chapter 5, Palumbo *et al.* highlighted the value of the extensive training materials that were developed by the Monitoring for Environment and Security in Africa (MESA) programme in extending the benefit of the eStation and Fire Monitoring Tool (FMT). Importantly, training was available in both English and French, extending the reach to additional conservation practitioners.

Customised Tool Development and Visualisations

Several case studies described how they built or are building specific tools and portals to allow end-users to access and customise information. Investing in higher-level data products helps improve data accessibility and build capacity, as noted above. However, these products still require technical expertise if they are to be used properly and are rarely the item that end-users need. The unique circumstances of each conservation challenge and the level of experience that end-users have means that projects need to take additional steps to deliver usable information. For example, Jantz *et al.* (Chapter 4) illustrate how forest-cover and forest-loss maps are insufficient for mapping and monitoring chimpanzee habitat. Instead, additional processing to develop habitat maps resulted in information that can be used by local communities. The fire-alert tools described in Palumbo *et al.* (Chapter 5) extract ecologically relevant information for a user's area of interest and deliver it to stakeholders in way that accommodates limited internet connectivity in many sub-Saharan African countries. Hebblewhite *et al.* in Chapter 7 developed and are expanding online software for stakeholders, complete with sample code and explanatory information. The message here is clear to us. It is the tools that deliver the needed data in a usable form that are key. These tools are the element that enables the transition from research to decision support, and that increases uptake by end-users.

Visualisations are also another key component to reaching end-users. Escribano and Fernández (Chapter 6) faced a daunting challenge in conveying the complicated topic of ecosystem health and function to managers in Doñana National Park. Their efforts to develop straightforward metrics to display resistance, resilience, and elasticity are particularly notable because they not only helped managers in the park understand the status and trends of the ecosystem, but gave them an appreciation for the role of remote sensing. Bailey *et al.* (Chapter 8) used team-member experience with a similar project, TurtleWatch, and feedback from end-users to refine visualisations displayed on the WhaleWatch project website. Rather than choosing maps based on biophysical variables, they invested in modelling efforts to display monthly whale occurrence and density information. Using such an approach provided concrete information to end-users who are looking to modify ship routes to avoid hitting blue whales. These two projects are particularly impressive because visually simple outputs emerged from complex analyses.

Integrating in situ *and Earth Observations*

As we note in our preface, satellite remote sensing is not a panacea for conservation monitoring. Even though satellites provide vital geographical coverage, repeated observations, and spectral information, which would be extremely costly or impossible to gather from field-based observations, none of the six case studies could have addressed conservation problems without *in situ* data. Long-term observations of shorebird declines identified the need for Murray (Chapter 3) to monitor tidal-flat loss, and the same ground-based observations will allow the International Union for Conservation of Nature to track the impact of tidal-flat protection on the status of key birds of conservation interest. Both Hebblewhite *et al.* (Chapter 7) and Bailey *et al.* (Chapter 8) relied on long time-series of mule deer and blue whale population data, respectively, in order to make optimal use of satellite remote sensing information. Thus, to maximise the benefit of Earth observations, the broader conservation community needs not only to invest in satellite systems, but also to maintain strong investments in collecting *in situ* data.

The conservation community will also benefit from sustained investments in basic and applied research to integrate *in situ* and remotely sensed data, an active and critical area of research for the biodiversity community (Pettorelli *et al.* 2014a). As more types of satellite observations become available and as the ground- and sea-based observations continue to diversify with the expansion of camera traps, audio recordings, and citizen science observations, there will be a huge potential for use-inspired research. Grant programmes at space agencies and other organisations, especially NASA, ESA, and the EC, have supported investments in integrating *in situ* and remotely sensed data, including many of the case studies highlighted in this book. Thus, ongoing funding of research and application activities (in addition to all of the activities mentioned above) is necessary to ensure that there will be new advances to take advantage of in the future to support conservation.

A Role for Serendipity

Several case studies highlighted the role of 'luck' in their projects, and we the editors have heard this phrase cited many times in the development of conservation projects. Hebblewhite *et al.* (Chapter 7) and Bailey *et al.* (Chapter 8) stated that they benefited from fortuitous meetings of key individuals or from having team members with helpful connections or experiences. While we certainly understand and agree that serendipity is important, we actually feel that we hear it so often that people downplay the underlying reason that many conservation projects have successful outcomes: hard work. Many scientists and conservationists have worked relentlessly to establish a strong reputation and spend time getting to know key people and organisations in their field, which serves as a basis for developing collaborative activities. Long-term study of an area, in-depth investigation as to the causes of conservation problems, investment in capacity-building, studying the cultural, political, and economic circumstances of a region, and a desire to see positive change are also key ingredients to conservation success. Thus, while we accept that 'luck' can contribute, we suspect that many of the case studies were simply aided by this ingredient and that the majority of success can be ascribed to hard work. It is important that individuals recognise their own hard work and the effort of others, so that we can measure how the conservation remote sensing community has progressed.

10.3 A LOOK TO THE FUTURE FOR CONSERVATION REMOTE SENSING

Observations of Earth from space have transformed our understanding of the physical and biological processes governing our planet. Satellite remote sensing has proven to be a powerful tool in biodiversity research and conservation applications, including operational biodiversity conservation. This is very timely. As outlined in Chapter 1, human pressures from over seven billion people living on Earth has resulted in the extinction of species, the decline of plant and animal populations, and the degradation and destruction of ecosystems. The subsequent loss of ecosystem function leads not only to declines of species, but also affects the ability of ecosystems to provide clean water, sequester carbon, and supply other services on which humans rely. Fortunately, we are able to harness human ingenuity and technology to develop and implement scientifically informed solutions to conservation challenges.

Since 2000, there has been a notable increase in the use of satellite remote sensing in biodiversity research and conservation applications, and there are many more opportunities to exploit currently available observations (Pettorelli et al. 2014b, Rose et al. 2015). New, passive optical observing systems will have higher spatial resolution than previous generations and will contribute to constellations that will result in a higher temporal resolution. Developments in hyperspectral technology might soon result in increases in spectral resolution too. Excitingly, there are also opportunities to take advantage of active instruments, such as radar and lidar that can provide information on the vertical structure of ecosystems. However, past experience suggests that there will be a lag before these new streams of information are incorporated into conservation operations. Data providers should not expect an immediate uptake of new products, even when space agencies and others make investments to generate higher-level products and develop tools to improve the ease of access and use. We anticipate that previous and ongoing investments in capacity-building will reduce a lag in uptake of new technology. If near-term capacity-building activities focus on preparing the community now for these new observations, assimilation into conservation applications can be expedited.

Satellite remote sensing now has a proven record for conservation. As with any discipline, conservation is susceptible to fads (Redford *et al.* 2013), and it would have been reasonable, 20 years ago, to question whether remote sensing was an overhyped topic that would fade in a few years. Had the observing technology not progressed, we might still be at that stage. Increased computing power, a move toward free and open data, and investments in data processing and access tools have also made a major contribution to the field (Turner *et al.* 2015). Through these advancements, and investments in capacity-building, a broad spectrum of conservationists has been exposed to the value of remote sensing. Although most of the studies in this book exploit data from observing systems that were available over a decade ago, there can be no doubt that the progressive advancement in sensors, especially their temporal and spatial resolution, also contribute to where we are now. The advances in the field specifically addressed many initial concerns in the community, two of the key ones being cost and spatial resolution.

An ongoing obstacle to expanding the use of satellite remote sensing in conservation remains a perceived scale mismatch. For those new to remote sensing, it is admittedly challenging to understand how data collected in tens, hundreds, and even thousands of metres can be useful to those thinking about site-specific or species-specific issues, where the scale of concern is an order of magnitude less than the remote sensing data. Turner *et al.* (2003) identified this impediment as a key technological and cultural limitation to the use of satellite remote sensing for conservation. As the case studies in this book show, observations at these spatial scales are very useful for conservation. Furthermore, the prevalent use of MODIS imagery (with a minimum resolution of 250-m) within the preceding local and regional case studies further solidifies the argument that the spatial resolution of data need not be a limiting factor.

We are pleased to see that strong collaborations between remote sensing scientists and conservationists have ensured, and continue to do so, that assumptions and biases in both communities do not limit ordinary or innovative applications. Perhaps the most important activity in harnessing remote sensing for conservation are these collaborations, as is illustrated by the six case studies presented in this book. Conservationists and remote sensing scientists are actively working together to grow the relatively new field of conservation remote sensing so that it blends capabilities of people trained in either or both disciplines. Observational continuity, free and open data, support for capacity-building, and investments in developing more usable Earth-observation data products were vital to the growth of the field. As the conservation remote sensing community continues to coalesce, we hope to see many more case studies presented in the future, especially those that fill in geographical gaps that we noted at the start of this chapter.

Based on our assessment of the case studies and our additional experiences in the field of conservation remote sensing, we feel that conservation projects that focus on developing customised tools and visualisations will provide the greatest benefit for expanding the use of remote sensing for operational conservation. End-user engagement will thus be critical to achieve successful tools and visualisations (indeed, this applies to any conservation project). Consequently, we encourage the communities of conservation scientists, conservation managers, and remote sensing scientists to work together and seek out new opportunities for collaboration.

Our intention with this book was to further inspire and aid the use of satellite remote sensing for conservation. As demonstrated in the case studies, it is critical to invest in free, open-access, pre-processed data products, which can then be further tailored to develop tools that address unique conservation circumstances. These advancements will require collaborative teams with diverse skills. Fortunately, assembling such teams is facilitated by the fact that conservation remote sensing is an exciting and expanding field that attracts people with a passion for science and a desire to conserve biodiversity on Earth. The need for satellite remote sensing applications is greater than ever as the biodiversity crisis deepens. Our community must continue to leverage as much as possible from national investments in satellite data that are made freely available. We encourage the community to keep moving forward and we sincerely look forward to updating this book in a decade, by which time we hope the prospects for global biodiversity are improving.

ACKNOWLEDGEMENTS

We thank all of the chapter authors for contributing to this book. Amanda Whitehurst provided invaluable feedback and encouragement on various drafts of this chapter, and Caitlin Gille assisted with editing.

REFERENCES

Burgman, M., Jarrad, F., and Main, E. (2015). Decreasing geographic bias in conservation biology. *Conservation Biology*, **29**, 1255–1256.

Claverie, M., Masek, J. G., Junchang, J., and Dungan, J. L. (2017). Harmonized Landsat-8 Sentinel-2 (HLS) product user's guide. Harmonized Landsat–Sentinel-2 (HLS) project. NASA. See https://hls.gsfc.nasa.gov.

CEOS (2017). 2017–2019 work plan: version 1.1.1. See http://ceos.org/document_management/Publications/CEOS_Work-Plans/CEOS_2017-2019-Work-Plan_Jul2017.pdf.

Green, R. E., Buchanan, G.M., and Almond, R. (2011). What do conservation practitioners want from remote sensing? Cambridge Conservation Initiative Report. See www.cambridgeconservation.org/resource/working-papers-and-reports/cci-report-what-do-conservation-practitioners-want-remote.

Lawler, J. J., Aukema, J. E., Grant, J. B., *et al.* (2006). Conservation science: a 20-year report card in a nutshell. *Frontiers in Ecology and the Environment*, **4**, 473–480.

Paganini, M., Leidner, A. K., Geller, G., Turner, W., and Wegmann, M. (2016). The role of space agencies in remotely sensed essential biodiversity variables. *Remote Sensing in Ecology and Conservation*, **2**, 132–140.

Pekel, J-F., Cottam, A., Gorelick, N., and Belward, A. S. (2016). High-resolution mapping of global surface water and its long-term changes. *Nature*, **540**, 418–422.

Pettorelli, N, Safi, K., and Turner, W. (2014a). Satellite remote sensing, biodiversity research, and conservation of the future. *Philosophical Transactions of the Royal Society B*, **69**, 20130190.

Pettorelli, N., Laurance, W. F., O'Brien, T. G., et al. (2014b). Satellite remote sensing for applied ecologists: opportunities and challenges. *Journal of Applied Ecology*, **51**, 839–848.

Redford, K. and Taber, A. (2000). Writing the wrongs: developing a safe-fail culture in conservation. *Conservation Biology*, **14**, 1567–1568.

Redford, K.H., Padoch, C., and Sunderland, T. (2013). Fads, funding, and forgetting in three decades of conservation. *Conservation Biology*, **27**, 1523–1739.

Rose, R. A., Byler, D., Eastman, J. R., et al. (2015). Ten ways remote sensing can contribute to conservation. *Conservation Biology*, **29**, 350–359.

Turner, W., Spector, S., Gardiner, N., et al. (2003). Remote sensing for biodiversity science and conservation. *Trends in Ecology & Evolution*, **18**, 306–314.

Turner, W., Rondinini, C., Pettorelli, N., et al. (2015). Free and open-access satellite data are key to biodiversity conservation. *Biological Conservation*, **182**, 173–176.

Index